Reviews of Environmental
Contamination and Toxicology

VOLUME 184

Reviews of Environmental Contamination and Toxicology

Continuation of Residue Reviews

Editor
George W. Ware

Editorial Board
Lilia A. Albert, Xalapa, Veracruz, Mexico
D.G. Crosby, Davis, California, USA · Pim de Voogt, Amsterdam, The Netherlands
O. Hutzinger, Bayreuth, Germany · James B. Knaak, Getzville, NY, USA
Foster L. Mayer, Gulf Breeze, Florida, USA · D.P. Morgan, Cedar Rapids, Iowa, USA
Douglas L. Park, Washington DC, USA · Ronald S. Tjeerdema, Davis, California, USA
David M. Whitacre, Summerfield, North Carolina · Raymond S.H. Yang, Fort Collins, Colorado, USA

Founding Editor
Francis A. Gunther

VOLUME 184

Coordinating Board of Editors

DR. GEORGE W. WARE, *Editor*
Reviews of Environmental Contamination and Toxicology

5794 E. Camino del Celador
Tucson, Arizona 85750, USA
(520) 299-3735 (phone and FAX)

DR. HERBERT N. NIGG, *Editor*
Bulletin of Environmental Contamination and Toxicology

University of Florida
700 Experimental Station Road
Lake Alfred, Florida 33850, USA
(941) 956-1151; FAX (941) 956-4631

DR. DANIEL R. DOERGE, *Editor*
Archives of Environmental Contamination and Toxicology

7719 12th Street
Paron, Arkansas 72122, USA
(501) 821-1147; FAX (501) 821-1146

Springer
New York: 233 Spring Street, New York, NY 10013, USA
Heidelberg: Postfach 10 52 80, 69042 Heidelberg, Germany

Library of Congress Catalog Card Number 62-18595.
Printed in the United States of America.

ISSN 0179-5953

Printed on acid-free paper.

© 2005 Springer Science+Business Media, Inc.
All rights reserved. This work may not be translated or copied in whole or in part without the written permission of the publisher (Springer Science+Business Media, Inc., 233 Spring St., New York, NY 10013, USA), except for brief excerpts in connection with reviews or scholarly analysis. Use in connection with any form of information storage and retrieval, electronic adaptation, computer software, or by similar or dissimilar methodology now known or hereafter developed is forbidden.
The use in this publication of trade names, trademarks, service marks, and similar terms, even if they are not identified as such, is not to be taken as an expression of opinion as to whether or not they are subject to proprietary rights.

Printed in the United States of America.

ISBN 0-387-22398-3   SPIN 10954799

springeronline.com

Foreword

International concern in scientific, industrial, and governmental communities over traces of xenobiotics in foods and in both abiotic and biotic environments has justified the present triumvirate of specialized publications in this field: comprehensive reviews, rapidly published research papers and progress reports, and archival documentations. These three international publications are integrated and scheduled to provide the coherency essential for nonduplicative and current progress in a field as dynamic and complex as environmental contamination and toxicology. This series is reserved exclusively for the diversified literature on "toxic" chemicals in our food, our feeds, our homes, recreational and working surroundings, our domestic animals, our wildlife and ourselves. Tremendous efforts worldwide have been mobilized to evaluate the nature, presence, magnitude, fate, and toxicology of the chemicals loosed upon the earth. Among the sequelae of this broad new emphasis is an undeniable need for an articulated set of authoritative publications, where one can find the latest important world literature produced by these emerging areas of science together with documentation of pertinent ancillary legislation.

Research directors and legislative or administrative advisers do not have the time to scan the escalating number of technical publications that may contain articles important to current responsibility. Rather, these individuals need the background provided by detailed reviews and the assurance that the latest information is made available to them, all with minimal literature searching. Similarly, the scientist assigned or attracted to a new problem is required to glean all literature pertinent to the task, to publish new developments or important new experimental details quickly, to inform others of findings that might alter their own efforts, and eventually to publish all his/her supporting data and conclusions for archival purposes.

In the fields of environmental contamination and toxicology, the sum of these concerns and responsibilities is decisively addressed by the uniform, encompassing, and timely publication format of the Springer-Verlag (Heidelberg and New York) triumvirate:

Reviews of Environmental Contamination and Toxicology [Vol. 1 through 97 (1962–1986) as Residue Reviews] for detailed review articles concerned with any aspects of chemical contaminants, including pesticides, in the total environment with toxicological considerations and consequences.

Bulletin of Environmental Contamination and Toxicology (Vol. 1 in 1966) for rapid publication of short reports of significant advances and discoveries in the fields of air, soil, water, and food contamination and pollution as well as methodology and other disciplines concerned with the introduction, presence, and effects of toxicants in the total environment.

Archives of Environmental Contamination and Toxicology (Vol.1 in 1973) for important complete articles emphasizing and describing original experimental or theoretical research work pertaining to the scientific aspects of chemical contaminants in the environment.

Manuscripts for *Reviews* and the *Archives* are in identical formats and are peer reviewed by scientists in the field for adequacy and value; manuscripts for the *Bulletin* are also reviewed, but are published by photo-offset from camera-ready copy to provide the latest results with minimum delay. The individual editors of these three publications comprise the joint Coordinating Board of Editors with referral within the Board of manuscripts submitted to one publication but deemed by major emphasis or length more suitable for one of the others.

Coordinating Board of Editors

Preface

Thanks to our news media, today's lay person may be familiar with such environmental topics as ozone depletion, global warming, greenhouse effect, nuclear and toxic waste disposal, massive marine oil spills, acid rain resulting from atmospheric SO_2 and NO_x, contamination of the marine commons, deforestation, radioactive leaks from nuclear power generators, free chlorine and CFC (chlorofluorocarbon) effects on the ozone layer, mad cow disease, pesticide residues in foods, green chemistry or green technology, volatile organic compounds (VOCs), hormone- or endocrine-disrupting chemicals, declining sperm counts, and immune system suppression by pesticides, just to cite a few. Some of the more current, and perhaps less familiar, additions include *xenobiotic transport, solute transport, Tiers 1 and 2, USEPA to cabinet status, and zero-discharge*. These are only the most prevalent topics of national interest. In more localized settings, residents are faced with leaking underground fuel tanks, movement of nitrates and industrial solvents into groundwater, air pollution and "stay-indoors" alerts in our major cities, radon seepage into homes, poor indoor air quality, chemical spills from overturned railroad tank cars, suspected health effects from living near high-voltage transmission lines, and food contamination by "flesh-eating" bacteria and other fungal or bacterial toxins.

It should then come as no surprise that the '90s generation is the first of mankind to have become afflicted with *chemophobia*, the pervasive and acute fear of chemicals.

There is abundant evidence, however, that virtually all organic chemicals are degraded or dissipated in our not-so-fragile environment, despite efforts by environmental ethicists and the media to persuade us otherwise. However, for most scientists involved in environmental contaminant reduction, there is indeed room for improvement in all spheres.

Environmentalism is the newest global political force, resulting in the emergence of multi-national consortia to control pollution and the evolution of the environmental ethic. Will the new politics of the 21st century be a consortium of technologists and environmentalists or a progressive confrontation? These matters are of genuine concern to governmental agencies and legislative bodies around the world, for many serious chemical incidents have resulted from accidents and improper use.

For those who make the decisions about how our planet is managed, there is an ongoing need for continual surveillance and intelligent controls to avoid endangering the environment, the public health, and wildlife. Ensuring safety-in-use of the many chemicals involved in our highly industrialized culture is a dynamic challenge, for the old, established materials are continually being displaced by newly developed molecules more acceptable to federal and state regulatory agencies, public health officials, and environmentalists.

Adequate safety-in-use evaluations of all chemicals persistent in our air, foodstuffs, and drinking water are not simple matters, and they incorporate the judgments of many individuals highly trained in a variety of complex biological, chemical, food technological, medical, pharmacological, and toxicological disciplines.

Reviews of Environmental Contamination and Toxicology continues to serve as an integrating factor both in focusing attention on those matters requiring further study and in collating for variously trained readers current knowledge in specific important areas involved with chemical contaminants in the total environment. Previous volumes of *Reviews* illustrate these objectives.

Because manuscripts are published in the order in which they are received in final form, it may seem that some important aspects of analytical chemistry, bioaccumulation, biochemistry, human and animal medicine, legislation, pharmacology, physiology, regulation, and toxicology have been neglected at times. However, these apparent omissions are recognized, and pertinent manuscripts are in preparation. The field is so very large and the interests in it are so varied that the Editor and the Editorial Board earnestly solicit authors and suggestions of underrepresented topics to make this international book series yet more useful and worthwhile.

Reviews of Environmental Contamination and·Toxicology attempts to provide concise, critical reviews of timely advances, philosophy, and significant areas of accomplished or needed endeavor in the total field of xenobiotics in any segment of the environment, as well as toxicological implications. These reviews can be either general or specific, but properly they may lie in the domains of analytical chemistry and its methodology, biochemistry, human and animal medicine, legislation, pharmacology, physiology, regulation, and toxicology. Certain affairs in food technology concerned specifically with pesticide and other food-additive problems are also appropriate subjects.

Justification for the preparation of any review for this book series is that it deals with some aspect of the many real problems arising from the presence of any foreign chemical in our surroundings. Thus, manuscripts may encompass case studies from any country. Added plant or animal pest-control chemicals or their metabolites that may persist into food and animal feeds are within this scope. Food additives (substances deliberately added to foods for flavor, odor, appearance, and preservation, as well as those inadvertently added during manufacture, packing, distribution, and storage) are also considered suitable review material. Additionally, chemical contamination in any manner of air, water, soil, or plant or animal life is within these objectives and their purview.

Normally, manuscripts are contributed by invitation, but suggested topics are welcome. Preliminary communication with the Editor is recommended before volunteered review manuscripts are submitted.

Tucson, Arizona   G.W.W.

Table of Contents

Foreword v
Preface vii

Organohalogen Contaminants in Delphinoid Cetaceans 1
    MAGALI HOUDE, PAUL F. HOEKSTRA, KEITH R. SOLOMON,
    AND DEREK C.G. MUIR

Environmental Contamination and Human Exposure to Lead
in Brazil 59
    MONICA M.B. PAOLIELLO AND EDUARDO M. DE CAPITANI

Arsenic Speciation and Toxicity in Biological Systems 97
    KAZI FARZANA AKTER, GARY OWENS, DAVID E. DAVEY,
    AND RAVI NAIDU

Organohalogen Contaminants in Delphinoid Cetaceans

Magali Houde, Paul F. Hoekstra, Keith R. Solomon, and Derek C.G. Muir

Contents

I. Introduction	1
II. Persistent Organohalogen Contaminants	2
A. Physicochemical Properties	2
B. Global Transport	3
C. Sources and Spatial Distribution	4
III. Marine Mammals and PHCs	5
A. Biotransformation	6
B. Pattern of Accumulation	6
C. Mechanisms of Action	7
D. Impacts on Physiological Systems	9
IV. Studying PHCs in Delphinoids: Methods and Results	11
A. Stranded, By-Catch, and Hunted Animals	11
B. Free-Ranging Delphinoid Populations	35
IV. Conclusions	40
Summary	41
Acknowledgments	41
References	41

I. Introduction

For centuries, marine mammals have been the cornerstone of industrial and economic activities in many countries, as well as an important part of the subsistence harvesting of several aboriginal populations. As a consequence of human-related activities such as commercial whaling, habitat degradation, and declining fish stocks, as well as accidental capture in fishing gear and physical and acoustical disturbance by ship traffic, many populations of marine mammals have been depleted compared to historic population estimates (Reeves et al. 2003).

An additional threat to the health of marine mammals is contamination by chemical pollutants. Assessment of chemical residues in marine mammal tissues

---

Communicated by George W. Ware.

M. Houde · K.R. Solomon · D.C.G. Muir
University of Guelph, Department of Environmental Biology, Guelph, Ontario, N1G 2W1, Canada

M. Houde · D.C.G. Muir (✉)
National Water Research Institute, Environment Canada, 867 Lakeshore Rd., Burlington, Ontario, L7R 4A6, Canada

P.F. Hoekstra
Golder Associates Ltd., Environmental Sciences Group, 2390 Argentina Road, Mississauga, Ontario, L5N 5Z7, Canada

has been conducted since the mid-1960s (Aguilar et al. 2002). The concern about the potential toxicity of chemical pollutants to physiological systems of marine mammal began when a series of die-offs occurred within a short period of time in different regions of the world. During the late 1980s, pinniped epizootics caused the deaths of approximately 20,000 northern European harbour seals (*Phoca vitulina*), hundreds of grey seals (*Halichoerus grypus*) (Osterhaus and Vedder 1988), and thousands of Baikal seals (*Phoca sibirica*) from Lake Baikal, Russia (Grachev et al. 1989). Additionally, during this period, more than 7000 bottlenose dolphins (*Tursiops truncatus*) were washed ashore along the central and southern Atlantic coastline of the United States (Kuehl et al. 1991). At the beginning of the 1990s, massive die-offs of striped dolphins (*Stenella coeruleoalba*) from the Mediterranean Sea (Aguilar and Borrell 1994a; Aguilar and Raga 1993; Kannan et al. 1993a) and bottlenose dolphins from the Gulf of Mexico (Kuehl and Haebler 1995) also occurred. In 2000, thousands of Caspian seals (*Phoca caspica*) from the Caspian Sea were found dead during another epizootic (Kennedy et al. 2000). Although deaths from die-offs were primarily attributed to distemper virus (or morbillivirus) infection (Aguilar and Raga 1993; Heide-Jørgensen et al. 1992; Kennedy et al. 2000), chemical contaminants have been suggested as contributing factors in these marine mammal epizootics.

Characterizations of chemical contaminants in tissues of Atlantic bottlenose dolphins that died during the American epizootic in 1987 reported significant concentrations of polychlorinated biphenyls (PCBs) (Kuehl et al. 1991). Likewise, assessment of environmental contaminants in the blubber of Mediterranean Sea dolphins indicated concentrations of PCBs two- to three times higher in striped dolphins found dead during the outbreak compared to free-ranging individuals sampled before and after the epizootic (Aguilar and Borrell 1994a). The high concentration of PCBs found in victims of mass mortalities, in addition to the known increased susceptibility to viral infections in experimentally exposed laboratory mammals such as mice, ducks, guinea pigs, and rhesus monkeys to PCBs (Friend and Trainer 1970; Thomas and Hinsdill 1978; Vos and De Roij 1972), led to the speculation about a possible suppression of immune functions associated with environmental pollution.

Few contaminant studies have been conducted on cetaceans, compared to pinnipeds, because of logistical, ethical, and legal constraints. The order *Cetacea* is divided into two suborders: Mysticeti, which are the baleen whales, and Odontoceti, which are toothed whales. In cetaceans, most chemical contaminant studies have been conducted on members of the toothed whale superfamily Delphinoidea (including three families: *Delphinidae*, *Phocoenidae* and *Monodontidae*), and these are reviewed here.

II. Persistent Organohalogen Contaminants
A. Physicochemical Properties

Exposure assessment of marine mammals to environmental contaminants is an important part of the management and conservation of wildlife populations. Persistent organohalogen contaminants (PHCs) are a major part of the investigation of marine pollution. Industrial chemicals such as PCBs and various organochlorine pesticides such as dichlorodiphenyltrichloroethane (DDT), chlordanes (CHL), hexachlorocyclohexanes (HCH), and hexachlorobenzene (HCB) are commonly studied PHCs in marine mammals. Other persistent pollutants, including tris(4-chlorophenyl)methane (TCPM) and tris(4-chlorophenyl)methanol (TCPMe), which may originate from the production of synthetic high polymers and light-fast dyes for acrylic fibers (Jarman et al. 1992), the insecticide toxaphene (Vetter et al. 2001a), perfluorinated acids (PFAs) such as perfluorooctane sulfonates (PFOS) and related compounds (Giesy and Kannan 2002), and polybrominated flame retardants, such as polybrominated diphenyl ethers (PBDEs) (McDonald 2002), are of concern because of their widespread distribution in marine ecosystems and their potential toxicity. Additionally, compounds from natural origins, such as halogenated dimethyl bipyrroles, have been globally detected in marine mammals (Tittlemier et al. 2002). Brominated and chlorinated dimethyl bipyrroles appear to be produced in the northern Pacific by an unknown pathway or mechanism, and their spatial distribution shows a different pattern than that of anthropogenic organohalogen contaminants (Tittlemier et al. 2002).

PHCs vary in their chemical structures and physicochemical properties such as vapor pressure, octanol–water partition coefficient, and Henry's law constant. In general, PHCs are chemically stable, resistant to biological or abiotic chemical breakdown, and, with the exception of perfluorinated acids, are lipophilic. This combination of physicochemical properties results in the environmental persistence of PHCs and their global dispersion and bioaccumulation in lipid-rich tissues of marine mammals. Moreover, the food chain transfer of PHCs and other contaminants to toothed whales has resulted in concentrations at several orders of magnitude greater than in species feeding at lower trophic levels, such as fish and baleen whales, as a result of biomagnification of contaminants through the food chain (Hoekstra et al. 2003).

B. Global Transport

The dispersion of PHCs in the marine environment depends on the physicochemical properties of compounds (particularly vapor pressure and air–water partition coefficient). Less volatile compounds, such as DDT and highly chlorinated PCBs, are usually found close to their emission source, as compared to the highly volatile chlorobenzene and HCH isomers, which are more globally distributed (Iwata et al. 1993; Li et al. 2002).

Long-range transfer of PHCs that are released in the environment may occur through oceanic and riverine transport. Global transfer of volatile and semivolatile PHCs can also occur through the volatilization of those contaminants from terrestrial soils and marine waters into the atmosphere, with ultimate deposition in remote regions through dry deposition of particles, gas exchange, or via precipitation scavenging of rain and snow (Macdonald et al. 2000). As a consequence of this long-range transfer, known as the "grasshopper effect" or global distillation (Mackay and Wania 1995), considerable concentrations of PHCs have been found in the air, water, snow, and ice and in biological samples of marine organisms inhabiting polar regions (de Wit et al., 2004). The highest concentrations of HCH in seawater, for example, are found in the Arctic environment. The extent of degradation (e.g., via photolysis, hydrolysis, microorganisms, or free radicals) of contaminants in the atmosphere, in marine waters, or in the nearshore sediment varies with chemical structure. PCBs have been found to degrade in the atmosphere due to reaction with hydroxy radicals (Mandalakis et al. 2003), whereas HCHs have been found to be slowly degraded by microorganisms in ocean water (Harner et al. 2000; Li et al. 2002).

C. Sources and Spatial Distribution

In 2001, more than 120 countries signed the Stockholm Convention on Persistent Organic Pollutants, targeting 12 specific PHCs (i.e., aldrin, DDT, dieldrin, endrin, HCB, mirex, toxaphene, PCBs, polychlorinated dioxins, and furans, and chlordanes, including heptachlor) (UNEP 2001). This treaty, ratified in May 2004, bans 8 chlorinated pesticides, prohibits PCB production, and establish a long-term goal of eliminating DDT use. However, the treaty permits public health exceptions for the use of DDT in mosquito control to fight malaria in developing countries. Many industrialized nations have already banned the manufacture and use of DDT, which has resulted in a marked decrease of the global concentrations of this contaminant in marine ecosystems during the past 30 years (Aguilar et al. 2002). Unlike DDT, PCB concentrations have generally declined slowly or not at all over time in the marine ecosystem. The inadequate disposal, improper storage, accidental release, and ongoing use of PCBs in building materials and electrical equipment all contributed to the continuous input of these toxicants into the environment (Aguilar et al. 2002; Breivik et al. 2002a; Tanabe 1988).

To evaluate the spatial distribution of PHCs, a worldwide survey was conducted in 1989–1990 by Iwata et al. (1993) to assess PHC concentrations in air and surface water from various oceans. The survey indicated a higher level of atmospheric contamination in the Northern Hemisphere compared to the Southern Hemisphere (Iwata et al. 1993). These results may be explained by the greater number of industrialized countries found in the northern region of the globe and relatively greater releases from more intense industrial activities as well as indirect release from commercial use. Breivik et al. (2002b) suggested that approximately 97% of the global historical use of PCBs has occurred in the Northern Hemisphere. As well, the most elevated concentrations of DDT were found in the surface water of tropical Asia (Iwata et al. 1993) and may be attributed to its ongoing use in these regions (Prudente et al. 1997) and its decreased mobility compared to more volatile PHCs such as HCB and α-HCH. Global soils surveys have shown that highest and lowest concentrations of PCBs were found in Europe and Greenland, respectively (Meijer et al. 2003). These observations were supported by Connell et al. (1999) in a systematic comparison of liphophilic chemicals in air, water, sediment, and biota of the Northern and Southern Hemispheres. More specifically, in North America, lowest concentrations of PCBs and DDTs in marine biota were found in Alaska and the western part of the Canadian Arctic compared to the eastern region (de March et al. 1998). Similar spatial trends were observed in water and zooplankton from Alaska and the Canadian Arctic (Hoekstra et al. 2002) as well as ringed seals (*Phoca hispida*) from the circumpolar region (Muir et al. 2000).

HCH isomers are widely distributed in the Northern Hemisphere waters (Li et al. 2002). These compounds are retained in cold water, resulting in an accumulation of HCH in the arctic marine ecosystem (Macdonald et al. 2000). Atmospheric concentrations of HCH have decreased during the past 15 years following application bans and restricted use (Macdonald et al. 2000). However, the insecticide γ-HCH is still widely used in North America and Europe (Li et al. 1998). The volatile HCB originates from pesticide usage, manufacturing, and combustion (Bailey 2001). In a global survey of HCB in background surface soils, the highest concentrations of this contaminant were found in the Northern Hemisphere (Meijer et al. 2003), as seen for PCBs. However, HCB concentrations in North America and Northern Europe have declined during the past 35 years based on measured concentrations in the atmosphere, sediment cores, fish, and bird eggs (Bailey 2001).

Emerging contaminants, such as PFAs and PBDEs, are following different trends. PFAs have a carbon chain in which all hydrogens have been replaced by fluorine, which confers thermal and chemical stability to the molecule. These compounds have been used since 1950 in a variety of household and industrial applications (e.g., surfactants, refrigerants, lubricants, paints, adhesives, paper coating) (Key et al. 1997). Perfluorinated acids have been found to be widely distributed in the environment and highly resistant to abiotic or metabolic degradation (Giesy and Kannan 2001, 2002). PBDE flame retardants are used in a wide range of manufactured products (e.g., computers, television sets, electrical cables), cars, and textile coatings (Darnerud et al. 2001). These lipophilic compounds have chemical structures very similar to PCBs and bioaccumulate in tissues (Haglund et al. 1997). Numerous studies show that PBDE concentrations are increasing in the environment (Darnerud et al. 2001; Ikonomou et al. 2002; Law et al. 2003a).

III. Marine Mammals and PHCs

The accumulation of persistent and bioaccumulative contaminants in the marine environment may impair the health of aquatic wildlife such as marine mammals. Many species of delphinoids experience lifelong exposure to persistent contaminants because of their elevated trophic position, longevity, and proximity of their coastal habitats to industrial and agricultural areas. As a result of trophic transfer of PHCs, odontocetes are exposed to greater amounts of contaminants through their dietary exposure (Morris et al. 1989) compared to Mysticetes, which feed at a lower trophic level (Hoekstra et al. 2003). Studies on lipid composition of striped dolphin tissues demonstrated that concentrations of

DDTs, HCHs, and highly chlorinated PCBs were correlated to triglycerides content mainly found in the blubber (Kawai et al. 1988). In general, cetaceans have large and dynamic lipid reserves for thermoregulation, energy storage, and buoyancy, which correspond also to a large accumulation capacity for lipophilic chemicals. Fat reserves may be mobilized in times of illness, starvation, pregnancy, and lactation, consequently increasing the potential bioavailability of lipophilic contaminants (Aguilar 1987).

A. Biotransformation

In mammals, xenobiotics circulating in the bloodstream are chemically transformed into more hydrophilic by-products by hepatic biotransformation processes to facilitate their excretion from the organism. The hydrophobicity of chlorinated xenobiotics, such as PCBs, depends on the degree and pattern of chlorination of the molecule. Lower chlorinated substances have structural features, such as adjacent unsubstituted carbon atoms, that make them generally less resistant to biotransformation compared to highly chlorinated ones. Biotransformation, however, can also result in the formation of more persistent and (or) biologically active metabolites, such as the conversion of p,p'-DDT to p,p'-DDE or the conversion of certain PCBs to hydroxylated and methyl sulfone ($MeSO_2$) PCBs (Letcher et al. 2000a).

Biotransformation of xenobiotics occurs principally in the liver and is divided into two phases. Phase I is responsible for the insertion of a polar group into the molecule, most often by the oxygen-mediated actions of cytochrome P450 (CYP450) or by removal of a chlorine molecule by dehalogenation (Parkinson 1996). Families of microsomal CYP450 play key roles in the detoxification or activation of xenobiotics in addition to their essential roles in the biosynthesis and catabolism of endogenous substrates such as steroid hormones, bile acids, fatty acids, and liposoluble vitamins (Honkakoski and Negishi 2000). Phase II of biotransformation consists of the conjugation of the xenobiotic functional groups present or introduced/exposed during phase I with endogenous compounds such as glutathione, amino acids, glucuronide, or sulfate (Parkinson 1996). Most of these reactions result in the increased hydrophilicity of the xenobiotics and therefore enhance their elimination. However, certain metabolic compounds such as $MeSO_2$-PCBs and $MeSO_2$-DDEs are sufficiently lipophilic to bioaccumulate in tissues (Letcher et al. 1998, 2000a). Other PHCs such as PBDEs have been shown to biotransformed to hydroxylated PBDE (HO-PBDE) metabolites in studies with laboratory rodents and fish, but there is presently no published information on PBDE metabolism in any cetacean species or population (Hakk and Letcher 2003). One recent exception was a report of hydroxylated PCBs (HO-PCBs) and HO-PBDEs in the plasma of killer whale (*Orcinus orca*) fed a diet of wild Pacific salmon (Bennett et al. 2002).

B. Pattern of Accumulation

In marine mammals, the uptake of PHCs from the diet exceeds the rate of metabolism and excretion. As a result, concentrations of lipophilic PHCs tend to increase with age in males and juvenile females (Cockcroft et al. 1989; Wells et al. 1994). When reaching maturity, females can transfer large quantities of pollutants to their offspring through gestation and during lactation via the lipid-rich milk (Addison and Brodie 1977; Borrell et al. 1995; Cockcroft et al. 1990), thus reducing their load of body contaminants. In long-finned pilot whales (*Globicephala melas*), 60% to 100% of the mother's contamination load was transferred by the lipid-rich milk to the offspring, compared to 4% to 10% through the placenta (Borrell et al. 1995). Maternal reproductive history is an important factor determining the contamination load transferred to the offspring. The firstborn calf receives a much higher burden of contaminants than subsequent offspring (Cockcroft et al. 1989; Ylitalo et al. 2001). Cockcroft et al. (1989) reported that 80% of the total contamination load of the mother was transferred to the firstborn in South African bottlenose dolphins. On the other hand, lipophilic PHCs continue to increase with age in adult males (Granby and Kinze 1991; Westgate et al. 1997), a phenomenon also observed in postreproductive female long-finned and short-finned pilot whales (*G. macrorhyncus*) (Tanabe et al. 1987a; Tilbury et al. 1999), as well as in postreproductive female killer whales (Ross et al. 2000). PCBs in male belugas (*Delphinapterus leucas*) have been shown to reach a plateau in older individuals (Stern et al. 1994) as they reach a near steady state with their environment.

C. Mechanisms of Action

Cytochrome P450 Induction Certain xenobiotics can stimulate the synthesis of enzymes involved in biotransformation processes, such as cytochrome P450. Three main inducible cytochrome P450 subfamilies (i.e., isozymes CYP1, CYP2, and CYP3) are known to be implicated in the biotransformation of marine pollutants (e.g., PCBs, chlorinated dioxins). CYP isoenzymes have different, sometimes overlapping, substrates. The cytochrome metabolic capacity varies with gender, age, tissue, and species as well as PHC exposure.

The CYP1A family catalyzes planar organics such as 2,3,7,8-tetrachlorodibenzo-*p*-dioxin (TCDD), which is considered as one of the most potent synthetic toxicants, and PCBs with no (or limited) *ortho*-, *meta*-substituted chlorine atoms on the biphenyl rings (i.e., dioxin-like PCBs). Induction of CYP1A enzyme synthesis occurs through the aryl hydrocarbon receptor (AhR). CYP2B and CYP3A catalyze reactions in molecules with no *meta*-, *para*-substituted chlorine (with at least one *ortho*-substituted chlorine) possessing a nonplanar configuration. Current knowledge of metabolic activity in cetaceans suggests that their metabolic capacity to degrade xenobiotics may be low compared to that of terrestrial mammals or pinnipeds (Boon et al. 1997; Tanabe et al. 1988, 1997a).

Biochemical and immunochemical analyses have revealed enzymes from the subfamily CYP1A in pinnipeds and cetaceans (Goksøyr 1995; Li et al. 2003; Teramitsu et al. 2000; Watanabe et al. 1989; White et al. 1994). Immunochemical assays have demonstrated the presence of CYP1A, CYP2B, and CYP3A isoenzymes in the liver of harbour seals (van Hezik et al. 2001). Moreover, AhR has recently been characterized in beluga whale liver tissue, suggesting that these animals are capable of metabolizing dioxin-like contaminants (Jensen and Hanh 2001).

In cetaceans, CYP1A has been identified in the liver of hunted and stranded belugas from the Canadian Arctic, and concentrations have been strongly correlated with non-*ortho* and mono-*ortho* PCB congeners in blubber (Muir et al. 1999a; White et al. 1994). White et al. (1994) found that CYP1A was a primary catalyst for EROD (7-ethoxy-*O*-deethylase) and AHH (aryl hydrocarbon hydroxylase) activities in Arctic belugas, with EROD activity levels falling within an overlapping range of concentrations measured in short-finned pilot whales, striped dolphins, and killer whales caught off the coast of Japan (Watanabe et al. 1989) and pilot whales stranded on Cape Cod (White et al. 2000). These results suggest that exposure of these delphinoids to high body burdens of contaminants, including PCBs, may be responsible for the induction of the CYP1A-like proteins (White et al. 1994, 2000). Based on PCB accumulation patterns, Muir et al. (1996a) suggested that belugas from the St. Lawrence estuary may have developed greater cytochrome activities (CYP1A/2B) in response to PCB exposure compared to less-contaminated arctic belugas. Similarly, Letcher et al. (2000b) reported a greater degrading capacity of PCB and DDE methyl sulfone precursors in free-ranging belugas from the St. Lawrence estuary compared to free-ranging belugas from western Hudson Bay, Canada. Letcher et al. (2000b) postulated that the differences of metabolic capacities between the two populations were associated with higher CYP enzyme activities in the southern St. Lawrence population.

The presence and activity of CYP1A suggest that PCB congeners with *ortho-meta*-unsubstituted chlorine atoms, in combination with at most one *ortho*-chlorine atom, can be metabolized by cetaceans (Boon et al. 1997; Reijnders 1994), albeit in a limited fashion compared to other mammals. On the other hand, low hepatic catalytic biomarkers of CYP2B-like enzymes have been observed in arctic belugas and pilot whales (Watanabe et al. 1989; White et al. 1994, 2000). Indirect characterization of CYP450 was conducted by evaluating the pattern of tissue PHC residues relative to those found in the environment. Norstrom et al. (1992) found a much higher abundance of nonplanar PCB congeners in Canadian Arctic belugas and narwhals, which suggested a low activity for CYP2B-type metabolism compared to a higher metabolic activity for CYP1A. Letcher et al. (2000b) suggested that CYP2B-like-mediated biotransformation may be induced in beluga whales based on the formation of methyl sulfone metabolites of PCB and DDE in their tissues. Nonetheless, cetaceans seem to have lower CYP2B-like activity than seals (Goksøyr 1995). The characterization and activity of these enzymes have yet to be determined in cetaceans.

Less is known about cytochrome enzymes from the subfamily CYP3A, which are very active in testosterone and bile acid metabolism (Honkakoski and Negishi 2000). These enzymes have wide substrate specificity and are involved in drug and xenobiotic metabolism (Honkakoski and Negishi 2000). CYP3A enzymes have been identified in pinnipeds and some species of whales such as sperm whales (*Physeter macrocephalus*) (Boon et al. 2001). Immunochemical and catalytic analyses have recently identified CYP3A-type enzymes in hepatic microsomes of belugas from the Canadian Arctic and the St. Lawrence estuary (McKinney et al. 2004). It has been suggested that CYP3A may play a role in the metabolism of PCB congeners containing a minimum of hydrogen substitution on *ortho-meta* carbons (Li et al. 2003; Nyman et al. 2003). Inhibition studies conducted with harbour and grey seal liver microsomes suggest the involvement of CYP3A-like isoenzymes in the metabolism of toxaphene (van Hezik et al. 2001). Positive correlations were found in blubber of harp seal (*Phoca groenlandica*) from the Barents Sea between testosterone 6β-hydroxylation (activity associated with CYP3A) and toxaphene levels, suggesting an induction of CYP3A-like activity by toxaphene exposure (Wolkers et al. 2000). CYP3A activity has not yet been observed in delphinoids.

Endocrine Disruption Some PHCs can act as endocrine disrupters, altering the synthesis, storage, release, transport, elimination, or binding of specific hormones. For example, HO-PCBs can exert their action on thyroid and steroid hormones that are essential elements in the development of the brain and sex organs as well as other hormone-dependent processes. HO-PCB retained in blood have a similar molecule structure to the thyroid hormone thyroxine (T_4). These metabolites have a high affinity and can competitively bind with thyroid hormone transport protein such as transthyretin (TTR) as well as other hormone-associated protein such as uteroglobin (Hagmar et al. 2001; Meerts et al. 2002). This reaction may prevent the formation of the thyroxin–TTR complex, which can reduce the transfer of retinol (vitamin A) and thyroxin to the target organs and decrease the availability of progesterone. Some PBDEs have also been found to decrease the concentrations of thyroxin and vitamin A in rodents (Hallgren et al. 2001). In addition, methyl sulfone metabolites of DDE are identified as potent adrenocorticolytic agents (toxic to the adrenal gland) in mice and mink (Brandt et al. 1992; Jönsson et al. 1992, 1993). The consequences of these xenobiotic actions on the endocrine functions in cetaceans are still unknown.

D. Impacts on Physiological Systems

Experimental exposures of laboratory mammals to PCBs, PBDEs, and DDT-related compounds have resulted in the depletion of immune functions (both cell-mediated and humoral immunities) and in the alterations of reproductive functions in mice, rats, guinea pigs, and monkeys (Banerjee 1987; Banerjee et al. 1986; Thomas and Hinsdill 1978; Vos and De Roij 1972; Zhou et al. 2001). Adverse reproductive, metabolic, immunological, and endocrinological effects of exposure of wildlife populations to environmental contaminants were also observed in North American piscivorous bird populations (Elliot et al. 1996; Grasman et al. 1996).

In marine mammals, evidence of reproductive impairment correlated with bioaccumulative contaminants has been reported in California sea lions (*Zalo-

phus californianus), ringed, grey, and harbour seals from the Baltic Sea (Addison 1989; DeLong et al. 1970; Helle et al. 1976a,b; Reijnders 1986). Studies are very limited in cetaceans. Reproductive impairment has been associated with elevated concentration of chemical contaminants in beluga whales from the St. Lawrence estuary (Béland et al. 1993; Martineau et al. 1987). Studies conducted on captive bottlenose dolphins living in netted open water in U.S. Navy facilities suggested that females with high body burdens of DDT and PCBs had a higher rate of stillborn or newborn mortality than females with lower contaminant loads (Reddy et al. 2001). In male Dall's porpoises from the North Pacific, a reduction in testosterone concentrations in blood was correlated with greater concentrations of PCBs and DDE in blubber (Subramanian et al. 1987), suggesting an antiandrogenic effect of these contaminants or the more rapid breakdown of steroids by the induced enzyme activity of PHCs and their metabolites. In addition to alteration of endocrine and reproductive functions, numerous pathologies such as a high incidence of tumors in California sea lions (Gulland et al. 1996) and tumors and adrenal lesions in St. Lawrence belugas (Béland et al. 1993; De Guise et al. 1994; Lair et al. 1997; Martineau et al. 1994, 2002) have been reported.

Studies on cause-and-effect relationships, as opposed to correlative relationships, between PHCs and adverse physiological effects on marine mammals are difficult to conduct. The first demonstration of a causal relationship between the two parameters was elaborated in a semifield study on captive harbour seals at the end of the 1980s (Reijnders 1986), in which one group of seals was fed with PCB-contaminated fish from the Wadden Sea and the other group was fed with relatively uncontaminated fish from the northeast Atlantic. Results of this experiment showed reproductive failure in seals fed with the contaminated fish (Reijnders 1986). In addition, significantly lower concentrations of plasma retinol and thyroid hormones were observed in seals fed with contaminated fish (Brouwer et al. 1989).

Reproductive impairment was also observed in a second semifield study in which captive harbour seals were fed with contaminated fish from the Baltic Sea (de Swart et al. 1994). Alteration in white blood cell counts, suppression of natural killer cell activity, delayed-type hypersensitivity, and antibody responses to ovalbumin were all correlated with chronic exposure to PCBs (de Swart et al. 1994, 1995a, 1996; Ross et al. 1996). *In vitro* assays that were conducted in parallel to the semifield study demonstrated the depression of lymphocyte T response and antigen-specific lymphocyte proliferative response (de Swart et al. 1994, 1995b; Ross et al. 1996) in seals fed with contaminated fish. The suppression of such immune functions may increase the susceptibility of animals to opportunistic infectious diseases caused by bacteria, viruses, or parasites, thus potentially leading to their death. Cause of death and burden of PCBs in blubber were analyzed during a field study on stranded harbour porpoises (*Phocoena phocoena*). PCB concentrations were significantly higher in harbour porpoises that died of infectious diseases compared to animals that died from physical trauma (Jepson et al. 1999).

The semifield experimental design was very useful for investigating *in vivo* the effects of dietary exposure of pinnipeds to PCBs; however, maternal transfer of PHCs was not addressed. Long-term and perinatal exposures to PHCs have been conducted in experimental studies on female mink (*Mustela vison*). Mink exposed to different concentrations of PCB mixtures as well as PCBs and DDE metabolites through diet demonstrated reproductive failures (fewer implantations, reduced birthrate and kit survival, smaller offspring than control animals) (Bleavins et al. 1980; Heaton et al. 1995; Lund et al. 1999), and high incidence of fetal deaths (Bäcklin and Bergman 1992; Heaton et al. 1995), as well as decreased thyroid hormone concentrations in female adult plasma and increased hepatic activity in both kits and female adults (Lund et al. 1999). Exposure of adult mink to PCBs also resulted in reduced liver and pulmonary retinol content (Hakansson et al. 1992).

IV. Studying PHCs in Delphinoids: Methods and Results
A. Stranded, By-Catch, and Hunted Animals

Collection of biological samples from wild cetaceans is a difficult task and thus the number of samples is usually relatively small. Bodies of cetaceans that die at sea are sometimes washed ashore. Stranded carcasses in good condition or tissues of animals caught in gillnets or that are hunted may be used for contaminant analysis. The monitoring of stranded marine mammals enables the determination of PHC concentrations in different tissues (e.g., blubber, liver, brain, muscles, lungs). The necropsy can also be conducted to detect pathologies and help determine the probable cause of death. Although no direct cause–effect relationship of pathologies with contaminant loads can be deduced from such observations, spatial and temporal comparisons of contamination patterns can be elaborated between and within species.

Important physiological and environmental parameters, such as age, sex, reproductive status (e.g., lactating, first-time mother), nutritional state (e.g., time of the last meal), body condition, metabolism, and excretory capacity of the animal as well as the trophic level, habitat, migration pattern, and pollution source may affect the concentration of PHCs in tissues and need to be considered when comparing contaminant loads between individuals or species (Aguilar 1987).

PHCs concentrations (ΣPCBs, ΣDDTs, HCB, ΣCHLs, ΣHCHs, ΣPBDEs, and PFAs) found in blubber of stranded, by-catch, or hunted delphinids (dolphins, killer whales, and pilot whales), phocoenids (porpoises), and monodontids (belugas and narwhals) are summarized in Tables 1 to 7. Tables include data collected from peer-reviewed literature published between 1980 and 2004. Variations in the concentration of lipophilic contaminants can occur depending on the lipid content of tissues (Hebert and Keenleyside 1995; Kawai et al. 1988). Therefore, the data were standardized on a lipid basis to facilitate comparisons and to minimize error between lipid extraction protocols (Aguilar 1987). When data expressed by authors were on a fresh (wet) weight basis, the lipid concen-

Table 1. Mean Concentration and Standard Deviation (SD) ($\mu g\ g^{-1}$ Lipid Weight) of Selected PHCs in Blubber of Stranded, By-Catch, and Hunted Dolphins.

Species and Location	N, Sex	Age or class	Year	ΣPCBs	ΣDDTs	HCB	ΣCHLs	ΣHCHs	Reference
Cephalorhyncus heavisidii (Heaviside's dolphin)									
W. South Africa	5m	—	1984–1987	1.3 ± 0.8	4 ± 2.2	0.06 ± 0.04	—	—	de Kock et al. 1994[a]
	3f	—	1977–1985	0.4 ± 0.5	2.7 ± 3	0.05 ± 0.04	—	—	
Cephalorhyncus hectori (Hector's dolphin)									
New Zealand	5m	1–11	—	1.6[17]	—	—	—	—	Jones et al. 1999[a]
	2f	<1–8	—	1.1[17]	—	—	—	—	
Delphinus delphis (common dolphin)									
South Africa	42	j	—	6.9	5.7[3]	—	—	—	Cockcroft et al. 1990[a]
	17m	a	—	11.3	9.1[3]	—	—	—	
N.W. Atlantic	4m	—	1986–1988	36.5 ± 4	14.4 ± 9[b]	0.2 ± 0.008	5.7 ± 2[4]	0.09 ± 0.05[c]	Kuehl et al. 1991
W. South Africa	10m	—	1984–1986	8.1 ± 7	12.5 ± 10.3	0.1 ± 0.2	—	—	de Kock et al. 1994[a]
	7f	—	1984–1987	4 ± 2.5	2.8 ± 1.7	0.06 ± 0.03	—	—	
Australia	1	—	—	0.3	—	—	—	—	Kemper et al. 1994[a]
N.W. North Pacific	2m	a	1987	—	—	—	—	1.6[3]	Tanabe et al. 1996[a]
Oregon, USA	2m	j,a	1991–1995	5.3	46.7[b]	—	—	—	Hayteas and Duffield, 1997[a]
N.W. North Pacific	3m	a	1987	29.6	36.3[3]	0.3	5.8[5]	1.7[3]	Prudente et al. 1997
Irish Sea	1	—	1987–1989	2.8[5]	—	—	—	—	Troisi et al. 1998
New Zealand	2m	a	—	1.3[17]	—	—	—	—	Jones et al. 1999[a]
N.W. North Pacific	3m	a	1987	32.8	11.9[3]	0.2	2.1[5]	0.2[3]	Minh et al. 2000
Ireland	5m	1–11	1993	9 ± 6[7]	9.4 ± 6.8[4]	—	1.9[3]	0.09[2]	Smyth et al. 2000
	3f	4–10	1993	4.2 ± 5.9[7]	4 ± 10[4]	—	0.9[3]	0.08[2]	
Mediterranean Sea, Spain	8m	j	1992–1994	20.4 ± 7.7[20]	19.3 ± 6.6[4]	—	—	—	Borrell et al. 2001
	3m	a	1992–1994	88.3 ± 37.7[20]	118.7 ± 54.8[4]	—	—	—	
	9f	j	1992–1994	25.4 ± 17.9[20]	25.6 ± 22.9[4]	—	—	—	
	2f	a	1992–1994	22.2 ± 16.9[20]	17.4 ± 14.5[4]	—	—	—	
N.E. Australia	1f	j	1995	0.6[10]	0.6[3]	0.04	0.3[5]	—	Vetter et al. 2001b
E. Australia	1f	j	1995	1.1[25]	0.7[3]	0.05	—	—	Law et al. 2003b

Table 1. (Continued).

Species and Location	N, Sex	Age or class	Year	ΣPCBs	ΣDDTs	HCB	ΣCHLs	ΣHCHs	Reference
Grampus griseus (**Risso's dolphin**)									
W. South Africa	1m	—	1984	3.9	10.7	0.03	—	—	de Kock et al. 1994[a]
	1f	—	1986	0.6	1.8	0.04	—	—	
Italy	1m	a	1992	1017[55]	667[2]	—	—	—	Corsolini et al. 1995
	1f	j	1992	41.7[55]	10.8[2]	—	—	—	
British Columbia, Canada	1m	a	1988	5.7	16.7[4]	0.1	2.8[9]	0.2[3]	Jarman et al. 1996[d]
Oregon, USA	1m	j	1991–1995	26.6	63.6[b]	—	—	—	Hayteas and Duffield 1997[a]
Italy	1m	a	1990	480[30]	304[6]	0.9	—	—	Marsili and Focardi 1997[e]
Off Taiji, Japan	5m	a	1991	128	43[3]	0.2	12.8[5]	0.4[3]	Prudente et al. 1997
Irish Sea	1	—	1987–1989	7.3[5]	—	—	—	—	Troisi et al. 1998
Lagenodelphis hosei (**Fraser's dolphin**)									
Off Taiji, Japan	4m	a	1991	—	—	—	1[5]	0.4 ± 0.08[3]	Tanabe et al. 1996[a]
Mindanao Sea, Philippines	1m	a	1996	10.6	50.7[3]	0.07	1[5]	0.3[3]	Prudente et al. 1997
Off Kii peninsula, Japan	3m	a	1991	72.9	38.6[3]	0.2	4.3[5]	0.2[3]	Minh et al. 2000
Mindanao Sea	3m	a	1996	6.3	10.9[3]	0.06	0.5[5]	0.07[3]	
Lagenorhyncus acutus (**Atlantic white-sided dolphin**)									
N.W. Atlantic	2m	—	1989	57.9	16[b]	0.1	15.8[4]	0.06[c]	Kuehl et al. 1991
	1f	—	1989	34.5	5.5[b]	0.2	4.3[4]	0.04[c]	
Faroe Islands	8m	—	1987	42.7 ± 18	22.5 ± 10.3[3]	—	—	—	Borrell 1993
	5f	—	1987	25.3 ± 21.2	15 ± 14.4[3]	—	—	—	
Ireland	8m	1–17	1994	40.5 ± 30.6[19]	18.3 ± 11.1[3]	0.3 ± 0.1	5.1 ± 2.9[5]	0.2 ± 0.08[2]	McKenzie et al. 1997
	9f	1–15	1994	8.6 ± 8.1[19]	2.8 ± 3.2[3]	0.2 ± 0.2	1.3 ± 1.7[5]	0.1 ± 0.1[2]	
Scotland	3m	1–9	1991–1993	30.7 ± 19.4[19]	50.1 ± 36.6[3]	1.2 ± 0.2	12.6 ± 8.9[5]	0.3 ± 0.1[2]	
	2f	2–6	1992	7.1[19]	5.4[3]	0.2	3.5[5]	0.2[2]	
Irish Sea	1	—	1987–1989	15.5[5]	—	—	—	—	Troisi et al. 1998
N.W. Atlantic	6	—	1994–1996	29.6 ± 16.2	32.1 ± 21.9[6]	0.4 ± 0.4	10.4 ± 8.3[5]	0.5 ± 0.4[4]	Weisbrod et al. 2001

Mean ± SD ($\mu g \ g^{-1}$ l.w.)

Table 1. (Continued).

Species and Location	N, Sex	Age or class	Year	Mean ± SD (µg g^{-1} l.w.)						Reference
				ΣPCBs	ΣDDTs	HCB	ΣCHLs	ΣHCHs		
Lagenorhynchus albirostris (white-beaked dolphin)										
Newfoundland, Canada	9m	2–3	1982	60.6 ± 52.8^{49}	81.3 ± 82.7^{5}	1.9 ± 1	22.7 ± 19.1^{12}	1.3 ± 0.2^{3}		Muir et al. 1988
	13f	2–4	1982	30.8 ± 9.7^{49}	38.6 ± 25.1^{5}	1.3 ± 0.4	11.7 ± 3.9^{12}	1.2 ± 0.4^{3}		
Lagenorhynchus obliquidens (Pacific white-sided dolphin)										
Off Iki Island, Japan	5m	a	1981	53.7 ± 7.7	108 ± 24.3^{3}	—	—	19.4 ± 1.3^{3}		Tanabe et al. 1983[a]
Iki Island, Japan	5m	a	1981	75.7	—	—	—	—		Tanabe et al. 1987b[a]
N. North Pacific	3m	a	1991	—	—	—	—	3.5 ± 0.7^{3}		Tanabe et al. 1996[a]
N. North Pacific	3m	a	1991	26.1	23.9^{3}	0.3	4.9^{5}	2.8^{3}		Prudente et al. 1997
Lagenorhynchus obscurus (dusky dolphin)										
S.W. South Pacific	1m	a	1980	2	13.7^{3}	—	—	0.01^{3}		Tanabe et al. 1983[a]
W. South Africa	6m	—	1977–1987	3.8 ± 2.7	10 ± 9.7	0.09 ± 0.06	—	—		de Kock et al. 1994[a]
	6f	—	1977–1987	1.7 ± 1	3.4 ± 2.6	0.04 ± 0.03	—	—		
New Zealand	1m	a	—	1.4^{17}	—	—	—	—		Jones et al. 1999[a]
Lissodelphis borealis (northern right whale dolphin)										
N. North Pacific	2m	a	1991	—	—	—	—	1.9^{3}		Tanabe et al. 1996[a]
N. North Pacific	2m	a	1991	35.7	38.1^{3}	0.3	3.7^{5}	0.7^{3}		Minh et al. 2000
Lissodelphis peronii (southern right whale dolphin)										
New Zealand	1m	1	—	0.8^{17}	—	—	—	—		Jones et al. 1999[a]
Sotalia fluviatilis (marine tucuxi dolphin)										
Guanabara Bay, Brazil	1m	—	1996	12.8^{6}	—	—	—	—		da Silva et al. 2003[a]
	1f	—	1996	3.7^{6}	—	—	—	—		
Cananéia estuary, Brazil	4m	a	1996–1998	5.7 ± 2.8^{27}	72.3 ± 51.5^{6}	0.02 ± 0.007	0.03 ± 0.01^{2}	0.03 ± 0.02^{4}		Yogui et al. 2003
	5f	a	1996–2001	3.7 ± 3.8^{27}	6.8 ± 4^{6}	0.01 ± 0.01	0.02 ± 0.009^{2}	0.006 ± 0.004^{4}		
Sousa chinensis (hump-backed dolphin)										
Bay of Bengal, India	3m	—	1990–1991	1.6 ± 0.5	16.5 ± 4.2^{4}	0.004 ± 0.004	—	0.6 ± 0.5^{4}		Tanabe et al. 1993
Bay of Bengal	2m	a	1992	—	—	—	—	0.2^{3}		Tanabe et al. 1996[a]
Bay of Bengal	3m	a	1992	7.9	94.8^{3}	0.02	0.2^{5}	1^{3}		Prudente et al. 1997

Table 1. (Continued).

Species and Location	N, Sex	Age or class	Year	ΣPCBs	ΣDDTs	HCB	ΣCHLs	ΣHCHs	Reference
Hong Kong	7m	—	1993–1997	72.1 ± 38.4	138 ± 62.5^3	0.2 ± 0.2	1 ± 0.4	2.9 ± 2.3	Minh et al. 1999
	4f	—	1996–1997	24.7 ± 20.9	61 ± 65.4^3	0.08 ± 0.03	0.5 ± 0.5	0.6 ± 0.8	
Stenella attenuata (pantropical spotted dolphin)									
Australia	1	—	—	1.2	1.7	0.01	0.006^2	—	Kemper et al. 1994[a]
Stenella coeruleoalba (striped dolphin)									
Cardigan Bay, W. Wales	1	j	1988	32.1	73.1^3	1.1	—	0.4^3	Morris et al. 1989
Japan	8m	a	1978–1979	41.4	54.3	0.3	—	0.6	Loganathan et al. 1990[a]
Japan	8m	a	1986	40	52.9	0.2	—	0.4	Kannan et al. 1993a
W. Mediterranean Sea	9m	2–18	1990	1300	478^2	—	—	—	
	1f	17	1990	290	69^2	—	—	—	
Mediterranean Sea	72	—	1990	778	—	—	—	—	Aguilar and Borrell 1994a[f]
W. South Africa	1m	—	1986	1.6	1.1	0.04	—	—	de Kock et al. 1994[a]
	1f	—	1984	1.8	1.5	0.03	—	—	
Mediterranean Sea	13m, 17f	—	1990–1992	856 ± 569	—	—	—	—	Borrell et al. 1996
Off Sanriku, Japan	4m	a	1992	—	53.6^b	—	—	1.1 ± 0.9^3	Tanabe et al. 1996[a]
Oregon, USA	1	—	1991–1995	8.4	128.1^6	0.9	—	—	Hayteas and Duffield 1997[a]
Italy	33m	—	1988–94	215.3^{30}	81.5^6	0.8	—	—	Marsili and Focardi 1997[e]
	26f	—	1988–93	92.8^{30}	57.8^3	0.3	—	—	
Off Sanriku, Japan	5m	a	1992	60.9	—	—	9.2^5	1.4^3	Prudente et al. 1997
Aegean Sea	4	—	1991	21.5 ± 2.8^5	—	—	—	—	Troisi et al. 1998
Mediterranean Sea	2m,1f	—	1989–1990	84.8 ± 11.4^{37}	—	—	—	—	Reich et al. 1999
Off Sanriku, Japan	1m	a	1992	74	34^3	0.2	5.8^5	0.6^3	Minh et al. 2000
W. Mediterranean Sea	12	—	1990–1992	75.1 ± 14.1^{20}	61.1 ± 12.3^3	—	—	—	Troisi et al. 2001
Stenella longirostris (spinner dolphin)									
Bay of Bengal	3m	—	1990	1.3 ± 0.8	36.8 ± 13.1^4	0.03 ± 0.03	—	0.8 ± 0.5^4	Tanabe et al. 1993
	2f	—	1990–1991	1.3	41.7^4	0.02	—	1^4	
Bay of Bengal	2m	a	1990	—	—	—	—	0.2^3	Tanabe et al. 1996[a]

Table 1. (Continued).

Species and Location	N, Sex	Age or class	Year	ΣPCBs	ΣDDTs	HCB	ΣCHLs	ΣHCHs	Reference
E. Tropical Pacific	2m	a	1980–1982	—	—	—	—	0.007[3]	Tanabe et al. 1996[a]
E. Tropical Pacific	3m	a	1980–1982	1.7	2.6[3]	0.02	0.2[5]	0.02[3]	Prudente et al. 1997
Bay of Bengal	3m	a	1990	2.7	38.3[3]	0.02	0.2[5]	0.2[3]	
Mindanao Sea, Philippines	2m	a	1996	12.4	69.4[3]	0.1	1.5[5]	0.4[3]	Minh et al. 2000
Mindanao Sea	2m	a	1996	3.8	7.4[3]	0.06	0.6[5]	0.07[3]	
Bay of Bengal	3m	a	1990	3.3	27.2[3]	0.02	0.2[5]	0.2[3]	
Steno bredanensis (rough-toothed dolphin)									
Italy	1f	—	1991	24.5[30]	7.3[6]	0.2	—	—	Marsili and Focardi 1997[e]
Gulf of Mexico									
Florida, USA	6m	<1–10	1997	47.6[31]	22.1[6]	0.06	4.1[7]	0.02[3]	Struntz et al. 2004
	9f	<1–42	1997	25.9[31]	14.1[6]	0.05	3.7[7]	0.06[3]	
Tursiops truncatus (bottlenose dolphin)									
E. South Africa	52m	a,j	1980–1987	20.2 ± 16.1	28.7 ± 29.4[3]	—	—	—	Cockcroft et al. 1989[a]
	52f	a,j	1980–1987	11.4 ± 15.1	7.9 ± 11.1[3]	—	—	—	
Cardigan Bay, W. Wales	1f	<1	1988	760	391[3]	1.7	—	2.1[3]	Morris et al. 1989
N.W. Atlantic	3m	—	1987–1988	138 ± 114	38.6 ± 55.1[b]	0.04 ± 0.01	—	—	Kuehl et al. 1991
	9f	—	1987–1989	62.4 ± 75.3	7.5 ± 11.9[b]	0.04 ± 0.05	—	—	
Scotland	1m	—	—	16.5[7]	—	—	—	—	Wells and Echarri 1992
Bay of Bengal, India	2m	—	1990	1.2	9.2[4]	0.03	—	0.2[4]	Tanabe et al. 1993
	2f	—	1990–1991	0.8	14.7[4]	0.006	—	0.2[4]	
W. South Africa	5m	—	1976–1985	3.7 ± 4.7	6.5 ± 6.7	0.03 ± 0.02	—	—	de Kock et al. 1994[a]
	1f	—	1987	2.3	2.5	0.01	—	—	
Australia	6	—	—	0.09	—	—	—	—	Kemper et al. 1994[a]
Scotland	1m	4	1988	12.3[7]	—	1.1	8.4[5]	—	Wells et al. 1994
Adriatic Sea, Italy	5f	1–17	1988–1990	8.5 ± 6.3[7]	394 ± 440[2]	0.4 ± 0.2	3.3 ± 1.7[5]	—	Corsolini et al. 1995
	5m	j,a	1992	1192 ± 792[55]	138[2]	—	—	—	
	2f	a	1992	587[55]		—	—	—	

Table 1. (Continued).

Species and Location	N, Sex	Age or class	Year	ΣPCBs	ΣDDTs	HCB	ΣCHLs	ΣHCHs	Reference
Gulf of Mexico, USA	2m,3f	j	1990	49	13.4^2	3.4	5.5^3	—	Kuehl and Haebler 1995
	8m	a	1990	93	37.9^2	0.3	6.1^3	—	
	5f	a	1990–1991	7.2	3.7^2	0.3	0.7^3	—	
Cardigan Bay, W. Wales	1m	4	1989	372^{25}	66.4^3	0.3	—	—	Law et al. 1995
	1m	23	1991	46.3^g	54.1^b	0.3	—	—	
Gulf of Mexico	14m, 19f	a,j	—	36.1	15.3^6	0.5	3.9^7	0.1^4	Salata et al. 1995
Italy	5m	—	1987–1992	35.7^{30}	12^6	0.5	—	—	Marsili and Focardi 1997[e]
	2f	—	1988–1992	24^{30}	3.2^6	0.2	—	—	
S. China Sea	1m	30	1994	12.8^{63}	13.7^3	0.5	0.2^5	0.6^3	Parsons and Chan 2001
	2f	4–6	1994–1995	12.4^{63}	111.5^3	0.2	1.3^5	0.3^3	
N.E. Australia	2m	a	1996–1997	16.1^{10}	32.2^3	0.02	4.8^5	—	Vetter et al. 2001b
	2f	a	1995–1999	1.3^{10}	1.3^3	0.02	0.4^5	—	
E. Australia	2f	a,j	1995–1996	2^{25}	1.2^3	0.05	—	—	Law et al. 2003b
E. Italy	7m	a,j	1999–2000	30.4 ± 17.5	—	—	—	—	Storelli et al. 2003
	2f	a,j	1999–2000	41	—	—	—	—	

Note: Superscript numbers indicate the total number of compounds included in the sum.
PCB, polychlorinated biphenyls; DDT, dichlorodiphenyltrichloroethane; HCB, hexachlorobenzene; CHL, chlordanes; HCH, hexachlorocyclohexanes; m, male; f, female; j, juvenile; a, adult; [a]Concentration converted from wet weight to lipid weight, assuming 70% of lipid content; [b]p,p'-DDE only; [c]γ-HCH only; [d]Geometric means; [e]Data expressed in $\mu g\ g^{-1}$ dry weight; [f]Medians; [g]CB 153 only.

Table 2. Mean Concentration and Standard Deviation (SD) ($\mu g\ g^{-1}$ Lipid Weight) of Selected PHCs in Blubber of Stranded Killer Whales.

Species and Location	N, Sex	Age of Class	Year	ΣPCBs	ΣDDTs	HCB	ΣCHLs	ΣHCHs	Reference
Orcinus orca (killer whale)									
Pacific, Japan	1m	a	1986	410	—	—	—	—	Ono et al. 1987[a]
Pacific, Japan	2f	a	1986	355	—	—	—	—	
Pacific, Japan	1f	a	1982	229	—	—	—	—	Tanabe et al. 1987b[b]
Pacific, Japan	3	—	1986	529[8]	—	—	—	—	Kannan et al. 1989[b]
Australia	1	—	—	—	40.6	0.6	0.2[c]	—	Kemper et al. 1994[b]
British Columbia, Canada	5m, 1f	j,a	1986–1989	24.2	35.2[4]	0.5	9.2[10]	0.8[3]	Jarman et al. 1996[d]
United Kingdom	1f	a	1995	33.5[25]	50[3]	1.5	—	0.2[2]	Law et al. 1997
Oregon, USA	3m	j	1988–1997	209 ± 333	249 ± 261[e]	—	—	—	Hayteas and Duffield 2000[b]
	2f	j,a	1995–1996	280.7	724[e]	—	—	—	
Pseudorca crassidens (false killer whale)									
British Columbia	2m	a	1987–1989	44.1	1064[4]	0.4	15.4[14]	0.5[3]	Jarman et al. 1996[d]

Note: Superscript numbers indicate the total number of compounds included in the sum. m, male; f, female; j, juvenile; a, adult; [a]Authors did not specify if data are expressed in wet weight or lipid weight; [b]Concentration converted from wet weight to lipid weight assuming 70% of lipid content; [c]Heptachlor only; [d]Geometric means; [e]p-p'-DDE only.

Table 3. Mean Concentration and Standard Deviation (SD) ($\mu g\ g^{-1}$ Lipid Weight) of Selected PHCs in Blubber of Stranded and Hunted Pilot Whales.

Species and Location	N, Sex	Age or Class	Year	Mean ± SD ($\mu g\ g^{-1}$ l.w.) ΣPCBs	ΣDDTs	HCB	ΣCHLs	ΣHCHs	Reference
Globicephala melas, G. melaena (long-finned pilot whale)									
Newfoundland, Canada	5m		1980	12.4 ± 5.4[49]	16.4 ± 8.7[5]	0.4 ± 0.3	2.3 ± 1.3[12]	0.2 ± 0.1[3]	Muir et al. 1988
	9f		1980	5.6 ± 5.5[49]	7.6 ± 9.1[5]	0.2 ± 0.2	1.1 ± 0.8[12]	0.1 ± 0.1[3]	
Faroe Islands	52m		1987	48.8 ± 23.1	31.4 ± 19.2[3]	—	—	—	Borrell 1993
	159f		1987	26.3 ± 23.1	13.4 ± 16.4[3]	—	—	—	
Australia	9		—	<0.07	<1.4	<0.07	—	<0.07[a]	Kemper et al. 1994[b]
Nordragota, Faroe Islands	35		1986	19.8 ± 13.2	—	—	—	0.5 ± 0.2[a]	Simmonds et al. 1994
Leynar, Faroe Islands	15		1986	33.6 ± 19	—	—	—	0.4 ± 0.2[a]	
Faroe Islands	24m	j	1987	35.6 ± 12.1	27.8 ± 9.6[4]	—	—	—	Borrell et al. 1995
	6m	a	1987	42.4 ± 6.8	33.6 ± 5[4]	—	—	—	
	31f	j	1987	38.5 ± 16.3	27.5 ± 13.9[4]	—	—	—	
	69f	a	1987	12.9 ± 6.9	6.4 ± 5[4]	—	—	—	
N. Atlantic	7		—	11.3 ± 7.6[33]	11.1 ± 6.7[6]	0.3 ± 0.3	2.7[5]	—	Becker et al. 1997[b]
Italy	1m		1990	125[30]	58.1[6]	0.1	—	0.06[2]	Marsili and Focardi 1997[c]
Irish Sea	1		1987–89	10.3[5]	—	—	—	—	Troisi et al. 1998
Massachusetts, USA	6 m	a,j	1986–90	17.4 ± 3.9[17]	14.1 ± 2.2[6]	0.5 ± 0.08	4.4 ± 0.8[5]	—	Tilbury et al. 1999
	16f	a,j	1986–90	8.1 ± 1.5[17]	8 ± 1.5[6]	0.3 ± 0.06	2.1 ± 0.4[5]	—	
Faroe Islands	173	j	1997	3.7 ± 0.8[d]	16.6[6]	0.5 ± 0.1	4.3 ± 0.6[5]	—	Dam and Bloch 2000
	54m	a	1997	4.1 ± 1[d]	18.3[6]	0.4 ± 0.09	3.9 ± 1.2[5]	—	
	193f	a	1997	1.8 ± 0.6[d]	6.6[6]	0.2 ± 0.09	1.8 ± 0.5[5]	—	
N.W. Atlantic	16	a,j	1990–1996	7.6 ± 7.1[26]	18.3 ± 23.7[6]	0.3 ± 0.3	2.3 ± 2.2[5]	—	Weisbrod et al. 2000
N.E. Atlantic, USA	1f	a	1991	—	1.3	—	—	—	Lebeuf et al. 2001
G. macrorhyncus (short-finned pilot whale)									
Ayukawa, Japan	5m	5.5–28.5	1985	8.9 ± 2.4	20.6 ± 7.1[e]	—	—	—	Tanabe et al. 1987a[b]
	24f	5.5–36.5	1985	4.4 ± 4.6	10.1 ± 9.9[e]	—	—	—	

Note: Superscript numbers indicate the total number of compounds included in the sum.
m, male; f, female; j, juvenile; a, adult; [a]γ-HCH only; [b]Concentration converted from wet weight to lipid weight assuming 70% of lipid content; [c]Data expressed in $\mu g\ g^{-1}$ of dry weight; [d]CB 153 only; [e]p,p'-DDE only.

Table 4. Mean Concentration and Standard Deviation (SD) ($\mu g\ g^{-1}$ Lipid Weight) of Selected PHCs in Blubber of Stranded, By-Catch, and Hunted Phocoenids.

Species and Location	N, Sex	Age or Class	Year	ΣPCBs	ΣDDTs	HCB	ΣCHLs	ΣHCHs	Reference
Neophocoena phocaenoides (finless porpoise)									
Seto, inland sea, Japan	1m	a	1985	57.1	—	—	—	—	Tanabe et al. 1987b[a]
Seto, inland sea	1	—	1985	457[8]	—	—	—	—	Kannan et al. 1989[a]
Hong Kong	3m	—	1996	31.3 ± 29.5	115.2 ± 88.2[3]	0.1 ± 0.03	0.5 ± 0.4	0.6 ± 0.7	Minh et al. 1999
	4f	—	1993–1997	9.9 ± 15.6	32.1 ± 38.3[3]	0.1 ± 0.1	0.2 ± 0.4	0.2 ± 0.3	
Phocoenoides dalli (Dall's porpoise)									
N. North Pacific/Bering Sea	5m	a	1980–1982	9.1 ± 4.1	11.3 ± 5.7[3]	—	—	1.7 ± 0.8[3]	Tanabe et al. 1983[a]
N.W. North Pacific (*dalli*-type)	12m	a	1984	12.9 ± 5.5	15.7 ± 4.4[b]	—	—	—	Subramanian et al. 1987[a]
N.N. Pacific	4m	a	1985	17.1	—	—	—	—	Tanabe et al. 1987b[a]
	2f	a	1980	2.1	—	—	—	—	
N. N. Pacific/Bering Sea	3m	a	1980–1982	—	9.1[3]	—	2.8[5]	0.8[3]	Kawano et al. 1988
Bering Sea (*dalli*-type)	4m	a	—	7.6	12.9[b]	—	—	—	Subramanian et al. 1988[a]
	1f	a	—	2.1	3.6[b]	—	—	—	
N.W. North Pacific (*dalli*-type)	14m	a	—	17.1	17.1[b]	—	—	—	
	13f	a	—	4	5.9[b]	—	—	—	
(*truei*-type)	9m	a	—	22.9	27.1[b]	—	—	—	
	1f	a	—	5.6	9.1[b]	—	—	—	
N. North Pacific	5	—	1980–1985	12.3[8]	—	—	—	—	Kannan et al. 1989[a]
British Columbia	2m, 1f	a,j	1987–1988	4.6	5.6[4]	0.1	3.4[12]	1.7[3]	Jarman et al. 1996[c]
Bering Sea	3m	a	1985	—	—	—	—	1.3 ± 0.4[3]	Tanabe et al. 1996[a]
N.E. North Pacific	3m	a	1987	—	—	—	—	2.2 ± 0.5[3]	
N.W. North Pacific	2m	a	1987	—	—	—	—	2.9[3]	
Japan Sea	1m	a	1989	—	—	—	—	8.6[3]	
Oregon, USA	1	fetus	1991–1995	—	2.3[b]	—	—	—	Hayteas and Duffield 1997[a]
Bering Sea	3m	a	1985	17.1	12.2[3]	0.8	3.7[5]	1.1[3]	Prudente et al. 1997
Japan Sea	3m	a	1989	29.7	53.8[3]	1.4	5.9[5]	7.1[3]	

Table 4. (Continued).

Species and Location	N, Sex	Age or Class	Year	ΣPCBs	ΣDDTs	HCB	ΣCHLs	ΣHCHs	Reference
N.E. North Pacific	3m	a	1987	15.5	17.9^3	0.7	3.5^5	1.9^3	Prudente et al. 1997
N.W. North Pacific	3m	a	1987	25.9	23.5^3	1	5.7^5	2.2^3	
Bering Sea	2m	a	1985	14.9	7.5^3	0.7	3.5^5	0.9^3	Minh et al. 2000
N.E. North Pacific	4m	a	1987	21.8	7.7^3	0.4	3.1^5	1^3	
Japan Sea	3m	a	1989	33.3	20.7^3	0.7	3.8^5	2.5^3	
N.W. North Pacific (*truei*-type)	6m	5–9	1984	33.7	28.3^3	1.6	8^4	2.9^3	Kajiwara et al. 2002
	8m	1–5	1998–1999	6.7	3.3^3	0.6	1.6^4	0.9^3	
	12m	5–9	1998–1999	24.1	11.8^3	0.8	5.6^4	1.4^3	
	7f	1–7	1998–1999	7.3	4.8^3	0.9	2.3^4	1^3	
	10f	3–10	1998–1999	4.9	2.7^3	0.2	1.3^4	0.3^3	
Hokkaido, Japan (*dalli*-type)	3m	a	1995	12.6	8.2^3	0.5	2^4	1.6^3	Kajiwara et al. 2002
	6m	8–10	1998	30.8	26.9^3	0.7	6^4	2.4^3	
	10f	4–9	1998–1999	3.7	3^3	0.09	1^4	0.3^3	
	4f	4–5	1998–1999	12.8	11.6^3	0.7	2.4^4	1.6^3	
***Phocoena phocoena* (harbour porpoise)**									
Cardigan Bay, W. Wales	4	j,a	1988	65.4 ± 32.5	13.6 ± 8.1^3	0.5 ± 0.1	—	0.4 ± 0.1^3	Morris et al. 1989
North Sea	1f	a	—	42.3^7	11.2^3	1.5	—	2.1^3	Beck et al. 1990
Danish North Sea	10m	0–4	1986–1988	13 ± 9.7^7	18.1 ± 19.9^3	0.7 ± 0.6	—	0.8 ± 0.7^2	Granby and Kinze 1991
	16f	0–>8	1986–1988	14.6 ± 12.9^7	14.7 ± 13.6^3	0.7 ± 0.8	—	0.7 ± 0.6^2	
West Greenland	1m	6	1988	4.5^7	9.3^3	0.9	—	0.2^2	
	1f	7	1988	0.7^7	0.8^3	0.2	—	0.1^2	
Scotland	1m	—	1990	33.2^7	—	—	—	—	Wells and Echarri 1992
Faroe Islands	3m	—	1987–1988	13.4 ± 2.4	5.6 ± 0.7^3	—	—	—	Borrell 1993
	3f	—	1988	8.8 ± 1	3.8 ± 0.4^3	—	—	—	
Puck Bay, Poland	3f	j	1989–1990	34 ± 11.5	12.6 ± 1.8^4	0.6 ± 0.3	1.2 ± 0.4^5	1.1 ± 0.4^4	Kannan et al. 1993b

Table 4. (Continued).

Species and Location	N, Sex	Age or Class	Year	ΣPCBs	ΣDDTs	HCB	ΣCHLs	ΣHCHs	Reference
Puck Bay, Baltic Sea	3f	j	1989–1990	33.9	—	—	—	—	Falandysz et al. 1994
Great Britain	48m	0–12	1989–1992	26.1 ± 26^{25}	7 ± 6.8^3	0.5 ± 0.4	—	0.3 ± 0.2^2	Kuiken et al. 1994
	46f	0–14	1989–1992	17.8 ± 24.1^{25}	4.7 ± 6.5^3	0.3 ± 0.3	—	0.2 ± 0.1^2	
Scotland	27m	j,a	1989–1991	5.9 ± 4.5^7	—	0.4 ± 0.4	1.9 ± 1.2^5	—	Wells et al. 1994
	19f	j,a	1989–1991	2.7 ± 2.5^7	—	0.3 ± 0.2	1 ± 0.9^5	—	
Scandinavia	34m	0–8+	1987–1991	23.3^{47}	16.4^5	0.6	2.5^3	0.7^3	Kleivane et al. 1995
British Columbia	4m,3f	a,j	1987–1989	9.1	8.9^4	0.5	8^{12}	1.2^3	Jarman et al. 1996^c
California	1m,2f	a,j	1987–1988	13.7	20.6^4	0.5	7.1^{14}	1.1^3	
North Sea, Netherlands	10m	1–13.5	1990–1993	36.5 ± 19.7^{28}	—	—	—	—	van Scheppingen et al. 1996
	12f	<1–7	1990–1992	19.9 ± 12.1^{28}	—	—	—	—	
Greenland	1m	11	1989	7.9^{28}	—	—	—	—	
	3f	3–9	1989	2.8 ± 0.6^{28}	—	—	—	—	
North Sea	16	—	1988–1995	—	9.1^3	1.2	0.7^d	—	Vetter et al. 1996^a
W. Iceland	4	—	1988–1995	1.4^e	—	—	—	—	
N. Atlantic	5	—	—	21.1 ± 8.9^{33}	10.4 ± 4.1^6	0.7 ± 0.5	6.7^5	0.7^2	Becker et al. 1997^a
Oregon, USA	12	—	1991–1995	3	3.6^b	—	—	—	Hayteas and Duffield 1997^{a,c}
Black Sea, Turkey	8m	4–8	1993	25 ± 7.5	—	—	—	—	Tanabe et al. 1997a
	2f	7–10	1993	6.6	—	—	—	—	
Hokkaido, Japan	6m	1–10	1993	7.5 ± 5.4	5.4 ± 4.2^4	0.4 ± 0.2	1.1 ± 0.6^5	1 ± 0.2^3	Tanabe et al. 1997b
Black Sea, Turkey	25m	1–8	1993	18.6 ± 9.3	80 ± 49.9^4	0.5 ± 0.1	1 ± 0.6^5	11.8 ± 4.6^3	
	24f	<1–10	1993	14 ± 9.1	60.5 ± 35.1^4	0.5 ± 0.2	0.7 ± 0.4^5	8.9 ± 5.5^3	
Newfoundland, Canada	18m	a,j	1991	5.9 ± 2.8^{68}	4.6 ± 2.1^6	—	4.3 ± 2^{14}	0.4 ± 0.09^3	Westgate et al. 1997
	11f	a,j	1991	6.2 ± 4.9^{68}	3.5 ± 2.6^6	—	3.1 ± 2^{14}	0.4 ± 0.2^3	
Gulf of St Lawrence, Canada	31m	a,j	1989–1991	12 ± 6.1^{68}	7.9 ± 4.5^6	—	5.7 ± 2.7^{14}	0.6 ± 0.2^3	
	31f	a,j	1989–1991	8.1 ± 4.4^{68}	5.1 ± 3.1^6	—	3.5 ± 2.1^{14}	0.4 ± 0.2^3	
Bay of Fundy/Gulf of Maine	55m	a,j	1989–1991	19.5 ± 12.6^{68}	8.7 ± 4.1^6	—	6.9 ± 3.7^{14}	0.4 ± 0.2^3	
	53f	a,j	1989–1991	12.9 ± 5.4^{68}	6.3 ± 2.9^6	—	4.3 ± 2.2^{14}	0.4 ± 0.2^3	

Table 4. (Continued).

Species and Location	N, Sex	Age or Class	Year	ΣPCBs	ΣDDTs	HCB	ΣCHLs	ΣHCHs	Reference
Irish Sea	1	—	1987–1989	6.2[5]	—	—	—	—	Troisi et al. 1998
Baltic Sea	13m	j	1985–1993	16 ± 8[6]	15 ± 18[4]	—	—	—	Berggrena et al. 1999
	4m	a	1988–1989	46 ± 29[6]	116 ± 134[4]	—	—	—	
Kattegat-Skagerrak Seas	10m	j	1989–1990	11 ± 5[6]	20 ± 13[4]	—	—	—	
	7m	a	1988–1990	13 ± 5.2[6]	25 ± 20[4]	—	—	—	
	5m	a	1978–1981	40 ± 22[6]	98 ± 43[4]	—	—	—	
West Norway	8m	a	1988–1990	15 ± 11[6]	9.1 ± 7.4[4]	—	—	—	
Baltic Sea	10m, 8f	j	1993–1995	14.9[46]	6.7[3]	0.3	—	0.6[2]	Bruhn et al. 1999[f]
North Sea	3m, 8f	j	1994–1995	17[46]	1.8[3]	0.2	—	0.4[2]	
Arctic waters	2m, 2f	j	1995	1.3[46]	1.4[3]	0.1	—	0.06[2]	
England and Wales	34	—	1990–1996	13.6[25]	—	—	—	—	Jepson et al. 1999
	33		1990–1996	31.1[25]					
Sweden	3	3–7	1996	13.1 ± 7.4	3 ± 1.6[6]	—	—	—	Karlson et al. 2000
Hokkaido, Japan	3m	a	1993	10.3	12.5[3]	0.5	1.6[5]	1.7[3]	Minh et al. 2000
Ireland	6m	1–6	1993–1994	6.2 ± 2.8[7]	3.2 ± 1.1[4]	—	1.1[3]	0.2[2]	Smyth et al. 2000
	6f	2–4	1993–1994	8 ± 3.3[7]	4.7 ± 1.6[4]	—	1.8[3]	0.3[2]	
Phocoena spinipinnis (Burmeister's porpoise)									
N. Argentina	4m	1+–5+	1989–1990	3.9 ± 1.8	5.6 ± 3.4[4]	—	—	—	Corcuera et al. 1995
	4f	0+–12+	1989–1990	2.3 ± 1.8	2.2 ± 2.2[4]	—	—	—	

Note: Superscript numbers equal the total number of compounds included in the sum.
m, male; f, female; j, juvenile; a, adult; [a]Concentration converted from wet weight to lipid weight assuming 70% of lipid content; [b]p,p'-DDE only; [c]Geometric means; [d]γ-CHL only; [e]CB153 only; [f]Medians.

Table 5. Mean Concentration and Standard Deviation (SD) ($\mu g\ g^{-1}$ Lipid Weight) of selected PHCs in Blubber of Stranded, By-Catch and Hunted Monodontids.

Species and Location	N, Sex	Age or Class	Year	ΣPCBs	ΣDDTs	HCB	ΣCHLs	ΣHCHs	Reference
***Delphinapterus leucas* (beluga)**									
St. Lawrence estuary, Canada	2m	33–44+	1983–1984	68.4	93.7[6]	—	—	—	Massé et al. 1986
St. Lawrence estuary	15m	<1–27+	1982–1985	211 ± 97.9	102.5 ± 80.2[3]	—	—	—	Martineau et al. 1987[a]
	10f	1+–29+	1982–1985	117 ± 211	26.3 ± 36.8[3]	—	—	—	
E. Hudson Bay	8m	5–33	1984–1985	3.2 ± 0.6	2.6 ± 0.8	0.4 ± 0.2	2.2 ± 0.4[14]	0.2 ± 0.07[3]	Muir et al. 1990
	8f	3–20	1984–1985	1.4 ± 1	1.1 ± 0.9	0.2 ± 0.1	1 ± 0.7[14]	0.2 ± 0.04[3]	
	6m	12–20	1987	—	6.5 ± 0.8	0.3 ± 0.2	—	—	
	6f	8–24	1987	—	2.2 ± 2.3	0.2 ± 0.09	—	—	
Cumberland Sound	6m	0–16	1983	5.9 ± 0.3	8.1 ± 2.3	1.1 ± 0.2	2.8 ± 0.5[14]	0.5 ± 0.1[3]	
	6f	0–17	1983	1.3 ± 0.5	1.1 ± 0.6	0.2 ± 0.05	0.7 ± 0.2[14]	0.3 ± 0.07[3]	
W. Hudson Bay	4m	8–19	1986	4.7 ± 0.5	4.7 ± 0.3	0.9 ± 0.08	3.5 ± 0.4[14]	0.4 ± 0.1[3]	
	4f	5–15	1986	1.3 ± 1.4	1.2 ± 1.3	0.3 ± 0.2	1.2 ± 1.1[14]	0.2 ± 0.05[3]	
Beaufort Sea	10m	—	1983–1987	4.5 ± 1.1	3 ± 1.1	0.8 ± 0.2	2.4 ± 0.6[14]	0.3 ± 0.08[3]	
	2f	—	1983–1987	1.5	0.8	0.4	0.8[14]	0.2[3]	
Jones Sound	8m	1–7.5	1984	2.9 ± 0.6	2.2 ± 0.4	0.6 ± 0.2	2.1 ± 0.5[14]	0.2 ± 0.1[3]	
	7f	1.5–10.5	1984	2.8 ± 2.3	2.5 ± 1.9	0.5 ± 0.2	2.1 ± 1.3[14]	0.2 ± 0.09[3]	
St. Lawrence estuary	4m	4–23.5	1986–1987	87.3 ± 17.6	116 ± 37.6	1.5 ± 0.5	8.6 ± 0.7[14]	0.4 ± 0.1[3]	
	5f	2.5–29	1986–1987	43.1 ± 25.4	26.6 ± 20	0.7 ± 0.5	4.1 ± 2.3[14]	0.3 ± 0.1[3]	
Beaufort Sea	5	—	1983	3.7 ± 1.9	—	0.7 ± 0.3	—	—	Norstrom and Simon 1990
Cumberland Sound	8	11.8 ± 4.1	1983	3.3 ± 2.4	—	0.7 ± 0.6	—	—	
W. Hudson Bay	19	11.6 ± 4.4	1984	2.9 ± 1.9	—	0.6 ± 0.4	—	—	
St. Lawrence estuary	2	—	1984–1986	20.9	42.7[3]	—	—	—	Jarman et al. 1992
Canadian Arctic	6m	—	—	5.6 ± 1.3[b]	—	—	—	—	Ford et al. 1993
St. Lawrence estuary	1f	25	1988	22.6	16.6	—	—	—	Bergman et al. 1994
West Greenland	71m	0–22	1989–1990	7.7 ± 3.2	5.8 ± 3.6[4]	—	3.4 ± 1.5	—	Stern et al. 1994[c]
	67f	<1–18.5	1989–1990	5.3 ± 3.3	3.8 ± 2.8[4]	—	2.6 ± 1.6	—	

Table 5. (Continued).

Species and Location	N, Sex	Age or Class	Year	ΣPCBs	ΣDDTs	HCB	ΣCHLs	ΣHCHs	Reference
St. Lawrence estuary	15m	<1–31+	1988–1990	78.9^{74}	81.1^4	1.4	8.4^{11}	0.5^3	Muir et al. 1996a[d]
	21f	2.5–31+	1987–1990	29.6^{74}	17.5^4	0.4	3.6^{11}	0.3^3	
Newfoundland	2f	15+–25+	1989–1990	2.9^{74}	2.1^4	—	—	—	Muir et al. 1996a
St. Lawrence estuary	9m	4–28	1993–1994	79.2^{88}	47.6^6	1	11.2^{11}	0.4^3	Muir et al. 1996b[d]
	7f	10–33	1993–1994	61.1^{88}	32.4^6	0.7	8.9^{11}	0.3^3	
Arctic	12	—	—	5.2 ± 2^{33}	3.6 ± 1.6^6	1 ± 0.4	2.3^5	0.2^2	Becker et al. 1997[c]
Cook Inlet, AK	12	—	—	1.4 ± 0.7^{33}	1.5 ± 0.9^6	0.5 ± 0.3	0.4	0.1 ± 0.1^e	
Alaska	2m	6–13	1992	7.3	5	—	3.2	—	Wade et al. 1997
	4f	11–19	1992	1.4 ± 0.6	0.7 ± 0.4	—	0.6 ± 0.3	—	
St. Lawrence estuary	1	neonate	1991	17.6^{29}	2.2^3	—	1.4^5	0.2^4	Gauthier et al. 1998
Cook Inlet, AK	10m	8.5–10.5	1992–1996	0.8 ± 0.4^{17}	1.5 ± 0.8^6	0.2 ± 0.1	0.6 ± 0.3^8	0.2 ± 0.08	Krahn et al. 1999
	10f	2–15	1994–1997	0.4 ± 0.3^{17}	0.7 ± 0.5^6	0.2 ± 0.2	0.3 ± 0.3^8	0.2 ± 0.06	
E. Beaufort Sea, AK	2f	4.5–8	1989	1.2^{17}	1.4^6	0.6	1.4^8	0.4	
E. Chukchi Sea	11m	6.5–19	1990–1996	3 ± 0.5^{17}	4.2 ± 1.1^6	0.9 ± 0.1	2.8 ± 0.5^8	0.4 ± 0.9	
	8f	6–27.5	1990–1996	0.8 ± 0.6^{17}	1.1 ± 1^6	0.3 ± 0.3	0.9 ± 0.7^8	0.3 ± 0.1	
St. Lawrence estuary	5m	12–26	1997–1998	—	127^6	—	—	—	Lebeuf et al. 2001
	5f	21–31	1997	—	11.7^6	—	—	—	
E. Canadian Arctic	3m	1+–6+	1994	0.2 ± 0.1^8	—	—	—	—	Helm et al. 2002
	3f	6–20+	1994	0.07 ± 0.05^8	—	—	—	—	
Beaufort-Chukchi Seas	20	9–40	1999–2000	3.3 ± 0.3^{103}	2 ± 0.2^6	—	1.3 ± 0.1^7	0.2 ± 0.04^3	Hoekstra et al. 2003
Monodon monoceros (narwhal)									
Canadian Arctic	15m	a	1982–1983	6.2 ± 1.6	7.1 ± 2.1^4	0.7 ± 0.2	2.3 ± 0.5	0.2 ± 0.07	Muir et al. 1992
	6f	a	1982–1983	3.1 ± 2.1	2.9 ± 2.3^4	0.5 ± 0.4	1.6 ± 1.2	0.2 ± 0.06	
Canadian Arctic	11m	—	—	6.3 ± 2^b	—	—	—	—	Ford et al. 1993
	6f	—	—	3.2 ± 2.2^b	—	—	—	—	

Note: Superscript numbers indicate the total number of compounds included in the sum. m, male; f, female; j, juvenile; a, adult; [a]Missing lipid content data were replaced by 70%; [b]Ortho and mono-*ortho* PCBs; [c]Concentration converted from wet weight to lipid weight assuming 70% of lipid content; [d]Geometric means; [e]α-HCH only.

Table 6. Mean Concentration and Standard Deviation (SD) ($\mu g\ g^{-1}$ Wet Weight) of Perfluorinated Compounds in Liver and Muscle of Stranded Delphinoids.

					Mean ± SD ($\mu g\ g^{-1}$ w.w.)				
Species and Location	N, Sex	Age or Class	Year	Tissue	PFOS[a]	PFOSA[b]	PFOA[c]	PFHxS[d]	Reference
Delphinus delphis									
Italy	1f	—	1998	Liver	0.9	0.9	<0.04	<0.02	Kannan et al. 2002
				Muscle	0.08	0.1	<0.04	<0.02	
Globicephala melas									
Italy	1	j	1996	Liver	0.3	0.05	<0.04	<0.02	Kannan et al. 2002
				Muscle	0.05	0.05	<0.04	<0.02	
Lagenorhyncus acutus									
S. North Sea	2	—	1995–2000	Liver	<0.01–0.03	—	—	—	van de Vijver et al. 2003
	1	—	1995–2000	Kidney	0.02	—	—	—	
Lagenorhyncus albirostris									
S. North Sea	7	—	1995–2000	Liver	0.01	—	—	—	van de Vijver et al. 2003
	7	—	1995–2000	Kidney	0.09	—	—	—	
Phocoena phocoena									
S. North Sea	48		1995–2000	Liver	0.9	—	—	—	van de Vijver et al. 2003
	43		1995–2000	Kidney	<0.01–0.8	—	—	—	

Table 6. (Continued).

Species and Location	N, Sex	Age or Class	Year	Tissue	Mean ± SD ($\mu g\ g^{-1}$ w.w.)				Reference
					PFOS[a]	PFOSA[b]	PFOA[c]	PFHxS[d]	
Stenella coeruleoalba									
Mediterranean Sea	4	—	1990s	Liver	0.1	—	—	—	Giesy and Kannan 2001
Florida	2m	13–16	1994–1997	Liver	0.2	—	—	—	Kannan et al. 2001
Italy	1m,3f	—	1991	Liver	0.03 ± 0.01	<0.04	<0.07	<0.007	Kannan et al. 2002
S. North Sea	2	—	1995–2000	Liver	0.01	—	—	—	van de Vijver et al. 2003
	2	—	1995–2000	Kidney	<0.01	—	—	—	
Stenella clymene (short-snouted spinner dolphin)									
Gulf of Mexico, Florida	3f	3–>18	1995	Liver	0.1 ± 0.04	—	—	—	Kannan et al. 2001
Steno bredanensis									
Florida	2f	7–15	1995	Liver	0.05	—	—	—	Kannan et al. 2001
Tursiops truncatus									
Mediterranean Sea	5	—	1990s	Liver	0.3	—	—	—	Giesy and Kannan 2001
Florida	8m	1.5–18	1991–2000	Liver	0.6 ± 0.5	—	—	—	Kannan et al. 2001
	12f	3.5–50+	1991–1999	Liver	0.4 ± 0.3	—	—	—	
Italy	6	—	1991–1992	Liver	0.06 ± 0.4	0.06 ± 0.4	—	—	Kannan et al. 2002

[a]Perfluorooctanesulfonate (PFOS); [b]Perfluorooctanesulfonamide (PFOSA); [c]Perfluorooctanoate (PFOA); [d]Perfluorohexanesulfonate (PFHxS).

Table 7. Mean Concentration and Standard Deviation (SD) ($\mu g\ g^{-1}$ Lipid Weight) of Polybrominated Diphenyl Ethers (PBDEs) in Blubber and Liver of Stranded, By-Catch, and Hunted Delphinoids.

Species and Location	N, Sex	Age or Class	Year	Tissue	ΣPBDEs, Mean ± SD ($\mu g\ g^{-1}$ l.w.)	Reference
Delphinapterus leueas						
St. Lawrence estuary	28m	9+ – 29+	1988–1999	Blubber	0.3 ± 0.2[10]	Lebeuf et al.
	26f	10–31.5+	1988–1999	Blubber	0.3 + 0.3[10]	2004
Globicephala melas						
Vestmanna, Faroe Islands	13m	j	1996	Blubber	3.2[19]	Lindström
	8m	a	1996	Blubber	1.6[19]	et al. 1999
	4f	j	1996	Blubber	3[19]	
	19f	a	1996	Blubber	1.1[19]	
Hvannasund, Faroe Islands	9f	a	1994	Blubber	0.8[19]	
Lagenorhyncus albirostris						
Netherlands, North Sea	1	—	—	Blubber	11[3]	de Boer et al. 1998[a]
Phocoena phocoena						
North Sea	—	—	—	Blubber	1.9[6]	Boon et al.
	—	—	—	Liver	3.2[6]	2002
Belgium, North Sea	15m, 6f	a,j	1997–2000	Liver	2.3 ± 1.8[8]	Covaci et al. 2002
England/Wales	22m	0–15	1996–1999	Blubber	2.5 ± 2.2[14]	Law et al.
	38f	<1–14	1996–1999	Blubber	2.1 ± 1.9[14]	2002
Tursiops truncatus						
N.W. Atlantic	3f	a	1987	—	0.2 ± 0.02	Kuehl et al. 1991
Gulf of Mexico, USA	2m,3f	j	1990	Blubber	1.9	Kuehl and
	8m	a	1990	Blubber	3.1	Haebler
	5f	a	1990–1991	Blubber	0.2	1995

Note: Superscript numbers indicate the total number of compounds included in the sum of PBDEs.
[a]Concentration converted from wet weight to lipid weight assuming 70% of lipid content.

trations were calculated from raw data (when available) or given mean and standard deviation by dividing the fresh weight basis by the extracted lipid content. When the lipid content of blubber was absent, a standard of 70% was considered to be representative of mean blubber lipid content for odontocetes (based on Aguilar et al. 2002). Perfluorinated acids have a pattern of accumulation different from lipophilic contaminants as they are usually found in plasma and proteinaceous tissues (e.g., liver); therefore, data in Table 6 are expressed on a wet weight basis. Sample collection, analytical method, and the number of congeners or isomers of each sum (Σ) group of PHCs analyzed may vary from one study to the other. Thus, caution should be used when comparing results among studies.

Spatial Trends of PHCs in Stranded, Hunted, and By-Catch Animals In general, the greatest loads of ΣPCBs and ΣDDTs and related metabolites were found in species inhabiting the midlatitudes of industrialized Asia, North America and Southern Europe (Fig. 1), reflecting the areas where these chemical compounds have been intensively used. Highest concentrations of PCBs and DDTs were

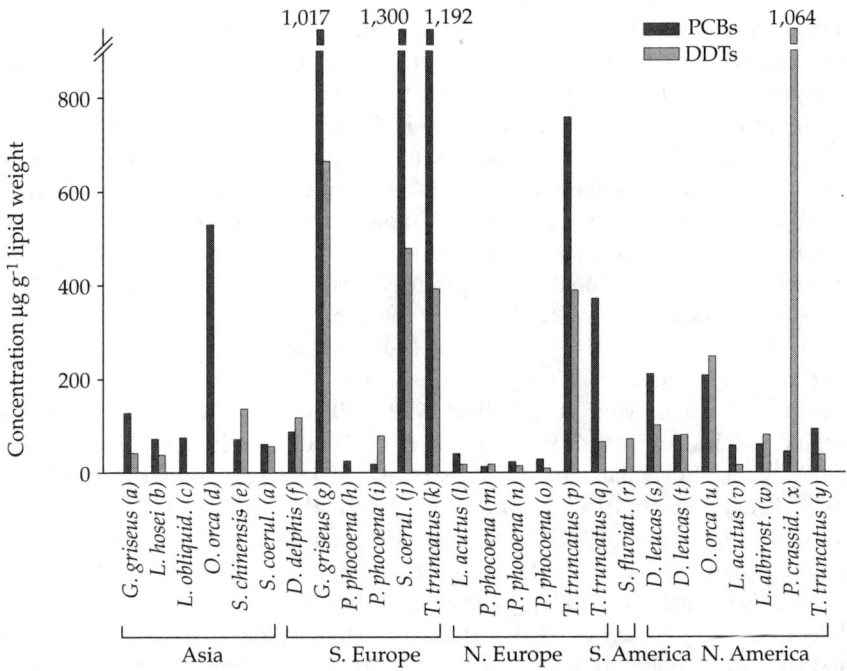

Fig. 1. Mean concentrations of polychlorinated biphenyls (ΣPCBs) and dichlorodiphenyl trichloroethane (ΣDDTs) [μg g^{-1} lipid weight] in blubber of stranded male delphinoids according to spatial locations. (a) Prudente et al. 1997; (b) Minh et al. 2000; (c) Tanabe et al. 1987b; (d) Kannan et al. 1989; (e) Minh et al. 1999; (f) Borrell et al. 2001; (g) Corsolini et al. 1995; (h) Tanabe et al. 1997a; (i) Tanabe et al. 1997b; (j) Kannan et al. 1993a; (k) Corsolini et al. 1995; (l) McKenzie et al. 1997; (m) Granby and Kinze 1991; (n) Kleivane et al. 1995; (o) Kuiken et al. 1994; (p) Morris et al. 1989; (q) Law et al. 1995; (r) Yogui et al. 2003; (s) Martineau et al. 1987; (t) Muir et al. 1996a; (u) Hayteas and Duffield 2000; (v) Kuehl et al. 1991; (w) Muir et al. 1988; (x) Jarman et al. 1996; (y) Kuehl and Haebler 1995.

detected in the blubber of offshore striped dolphins [mean PCB concentrations: 1300 µg g^{-1} lipid weight (l.w.), 856 µg g^{-1} l.w., and 778 µg g^{-1} l.w. (median); mean DDT concentrations: 480 µg g^{-1} l.w.] (Aguilar and Borrell 1994a; Borrell et al. 1996; Kannan et al. 1993a) and Risso's dolphins (*Grampus griseus*) from the Mediterranean Sea (PCBs: 1017 µg g^{-1} l.w. and 480 µg g^{-1} dry weight (d.w.); DDTs: 667 µg g^{-1} l.w. and 304 µg g^{-1} d.w.) (Corsolini et al. 1995; Marsili and Focardi 1997), bottlenose dolphins from Italy (PCBs: 1192 µg g^{-1} l.w.; DDTs: 394 µg g^{-1} l.w.) (Corsolini et al. 1995) and Britain (PCBs: 372 µg g^{-1} l.w.) (Law et al. 1995), as well as in killer whales from the Pacific coast of Japan (mean PCB concentrations 529 µg g^{-1} l.w.) (Kannan et al. 1989). Jarman et al. (1996) reported concentrations of the sum of four DDT congeners from two male adult false killer whales (*Pseudorca crassidens*) stranded in British Columbia of 1064 µg g^{-1} l.w. (geometric mean). A 10-mon-old female bottlenose dolphin calf from Cardigan Bay, West Wales, was found in 1988 with a ΣPCBs concentration of 760 µg g^{-1} l.w. (Morris et al. 1989). Transfer of the contamination load of the mother via milk may provide a likely explanation for the relatively high burden of PCBs in this calf.

Additionally, high concentrations of PCBs (ranging from 209 to 281 µg g^{-1} l.w.) and *p,p'*-DDE (ranging from 249 to 724 µg g^{-1} l.w.) were found in killer whales from the northwest coast of the United States (Hayteas and Duffield 2000) as well as in adult male Risso's dolphins from Japan (mean PCB concentration: 128 µg g^{-1} l.w.; mean DDT concentration: 43 µg g^{-1} l.w.) (Prudente et al. 1997). Other delphinoids from the Mediterranean Sea (Borrell et al. 2001; Marsili and Focardi 1997; Reich et al. 1999; Troisi et al. 2001), the U.S. Atlantic coast (Kuehl et al. 1991), the Gulf of Mexico (Kuehl and Haebler 1995), the east and west coasts of Canada (Jarman et al. 1996; Martineau et al. 1987; Massé et al. 1986; Muir et al. 1988, 1990, 1996a,b), and the coastal waters of Ireland (McKenzie et al. 1997; Smyth et al. 2000), Scotland (McKenzie et al. 1997; Wells and Echarri 1992; Wells et al. 1994), Great Britain (Jepson et al. 1999; Kuiken et al. 1994; Morris et al. 1989), Poland (Kannan et al. 1993b), Norway (Berggrena et al. 1999), Sweden (Berggrena et al. 1999), Turkey (Tanabe et al. 1997a,b), Scandinavia (Kleivane et al. 1995), and Hong Kong (Minh et al. 1999; Parsons and Chan 2001; Prudente et al. 1997) also show elevated concentrations (>15 µg g^{-1} l.w.) of PCB and DDT contamination. Lowest concentrations of PHCs in delphinoids were found in the blubber of beluga whales from the Arctic (Becker et al. 1997; Helm et al. 2002; Krahn et al. 1999; Muir et al. 1990, 1996a, 1999a; Stern et al. 1994). CHL concentrations were quite significant in some populations such as in male white-beaked dolphins (*Lagenorhyncus albirostris*) from the Gulf of St. Lawrence (22.7 µg g^{-1} l.w.) (Muir et al. 1988) and male white-sided dolphins (*Lagenorhyncus acutus*) from Scotland (12.6 µg g^{-1} l.w.) (McKenzie et al. 1997). The number of isomers and metabolites included in the ΣCHL may be a factor explaining differences in CHL concentrations. Other contaminants such as HCB and HCHs are also widely distributed worldwide, but their concentrations are generally far less prominent in delphinoids.

Concentrations of perfluorinated acids [i.e., perfluorooctane sulfonate (PFOS), perfluorohexane sulfonate (PFHxS) and perfluorooctanoate (PFOA)] and related compounds [i.e., perfluorooctanesulfonamide (PFOSA)] in cetaceans have been assessed in only a limited number of studies. Perfluorinated contaminants were found in tissues of stranded delphinoids inhabitating the Mediterranean and North Seas as well as Florida waters (Giesy and Kannan 2001; Kannan et al. 2001, 2002; van de Vijver et al. 2003) (see Table 6). Lowest concentrations of PFOS (<0.01 µg g^{-1} w.w.) were found in dolphin tissues from the North Sea (van de Vijver et al. 2003), and highest PFOS concentrations (0.9 µg g^{-1} w.w.) were detected in the liver of a common dolphin from the Mediterranean Sea (Kannan et al. 2002). PFOS and PFOA have also been detected in liver of belugas (concentrations ranging from 0.009–0.01 µg g^{-1} w.w.) and narwhals (*Monodon monoceros*) (concentrations ranging from 0.004–0.02 µg g^{-1} w.w.) from the eastern Canadian Arctic (Tomy and Helm 2003). Results from this study indicate biomagnification of PFOS and PFOA concentrations from prey to predators (i.e., fish to odontocetes; Tomy and Helm 2003).

PBDEs have been found in blubber and liver of delphinoids from different regions of the world. ΣPBDEs in blubber of harbour porpoises from England and Wales show a mean concentration of 2.5 µg g^{-1} l.w. in males and 2.1 µg g^{-1} l.w. in females, with the highest PBDE concentration (7.8 µg g^{-1} l.w.) found in a 2-year-old male porpoise (Law et al. 2002) (see Table 7). These results are similar to PBDE concentration (sum of three congeners) found in blubber of a single white-beaked dolphin found stranded on the Dutch coast of the North Sea (11 µg g^{-1} l.w.) (de Boer et al. 1998). Lower PBDE concentrations, ranging from 0.3 to 2.3 µg g^{-1} l.w., were observed in the blubber of 5 male harbour propoises from the west coast of Canada (Ikonomou et al. 2000) as well as in blubber of belugas from the Canadian Arctic (concentrations in males: 0.2 µg g^{-1} l.w.; females: 0.08 µg g^{-1} l.w.) (Alaee et al. 1999). Summed concentrations of 10 PBDEs measured in blubber of 54 adult beluga whales from the St. Lawrence estuary varied from 0.2 to 0.8 µg g^{-1} l.w. in males and from 0.3 to 1.1 µg g^{-1} l.w. in females (Lebeuf et al. 2004), accumulation in both genders showed significant exponential increase throughout the 1988–1999 time interval (Lebeuf et al. 2004). The lowest concentrations were found in blubber of beluga whales collected from Baffin Island, Canada in 1994 where mean PBDE concentrations in males and females were 0.063 and 0.002 µg g^{-1} l.w. respectively (Muir 1999b). Law et al. (2003a) have reported data on PBDEs in cetaceans and other marine mammals. PBDEs were detected in blubber of white-sided dolphin ($n = 1$; 0.3 µg g^{-1} l.w.), white-beaked dolphins ($n = 2$; 10.5 µg g^{-1} l.w.), striped dolphin ($n = 1$; 1.2 µg g^{-1} l.w.), common dolphin ($n = 1$; 0.5 µg g^{-1} l.w.), and long-finned pilot whale ($n = 1$; 1.7 µg g^{-1} l.w.) from England and Wales as well as in blubber of two bottlenose dolphins from Australia (1995–1996; 0.08 µg g^{-1} l.w. and 0.2 µg g^{-1} l.w.). Interestingly, the concentration of PBDEs found in a harbour porpoise fetus corresponded to 60% of the level found in its mother, suggesting possible transplacental transfer of these compounds (Law et al. 2002). Based on the high ΣPBDE concentrations found in immature male and female pilot whales compared to adults, Lindström et al. (1999) also suggested a lactational transfer of PBDEs from mother to offspring. van Bavel et al. (1999) reported sums of 21 PBDE congeners in the blubber of long-finned pilot whales from the Faroe Islands ranging from 0.1 to 1.3 µg g^{-1} l.w. Lowest concentrations were found in adult females and the highest in juvenile females.

Because of logistical and financial constraints, fewer investigations of PHCs in marine mammals have been conducted in the Southern Hemisphere. PHC analyses of delphinoid whales in this region show lower concentrations of contamination in marine mammal tissues compared to other locations (Aguilar et al. 2002; Kemper et al. 1994). Blubber PCB in stranded pilot whales from New Zealand ($n = 61$; mean concentration: 0.31 µg g^{-1} w.w.) were lower than those in hunted long-finned pilot whales from Faroe Islands ($n = 417$; mean concentration, 11.9 µg g^{-1} w.w.) (Dam and Boch 2000). However, PCBs, DDT, and dieldrin concentrations in blubber of dead bottlenose dolphins from industrialized regions of southern Australia were similar to those recorded in the highly contaminated regions of the Northern Hemisphere (Ruchel 2001).

Cetaceans inhabiting nonindustrialized regions in the Northern and Southern Hemispheres are not free of PHCs. In remote regions, concentrations of contaminants are usually lower, and patterns are unique compared to urbanized areas due to absence of local sources. Spatial comparison between beluga populations indicated more elevated concentrations of PHCs in the St. Lawrence estuary belugas compared to arctic populations (Andersen et al. 2001; Helm et al. 2002; Hobbs et al. 2003; Martineau et al. 1987; Massé et al. 1986; Muir et al. 1990, 1996a,b). Arctic delphinoids are less contaminated by PHCs than southern populations (see Tables 1–5) (Aguilar et al. 2002), which can be explained by the limited transport from the use areas. However, harbour porpoises from Arctic waters show greater concentrations of less chlorinated PCBs than individuals from the Baltic or North Seas (Bruhn et al. 1999), a pattern also observed in narwhals from Canada (Muir et al. 1992) and other odontocete species from the North Pacific Ocean (Prudente et al. 1997).

Extrapolating laboratory and surrogate species data to cetaceans is a difficult task. The absence of experimental studies in delphinoids restricts comparisons with other ecotoxicological studies. Assessment of PCBs in different delphinoid populations from the Northern Hemisphere shows concentrations above the thresholds for biological effects seen in mammals (Fig. 2). Unlike delphinoids from the Northern Hemisphere, PCBs in blubber of individuals from the Southern Hemisphere were generally lower than thresholds for mammalian adverse effects. Lipid-normalized PCBs in several populations exceed the threshold associated with vitamin A reduction in otters (11 µg g^{-1} l.w.) (Murk et al. 1998), and immune function and vitamin A disruption in harbour seals (16.5 µg g^{-1} l.w.) (de Swart et al. 1994, 1996). PCB concentrations known to reduce the litter size and the survival of kits of minks (40–60; 80–120 µg g^{-1} l.w.) (Leonards et al. 1995), as well as reproductive success in ringed seals (77 µg g^{-1} l.w.) (Helle et al. 1976a), were exceeded by those in many delphinoid populations from the Northern Hemi-

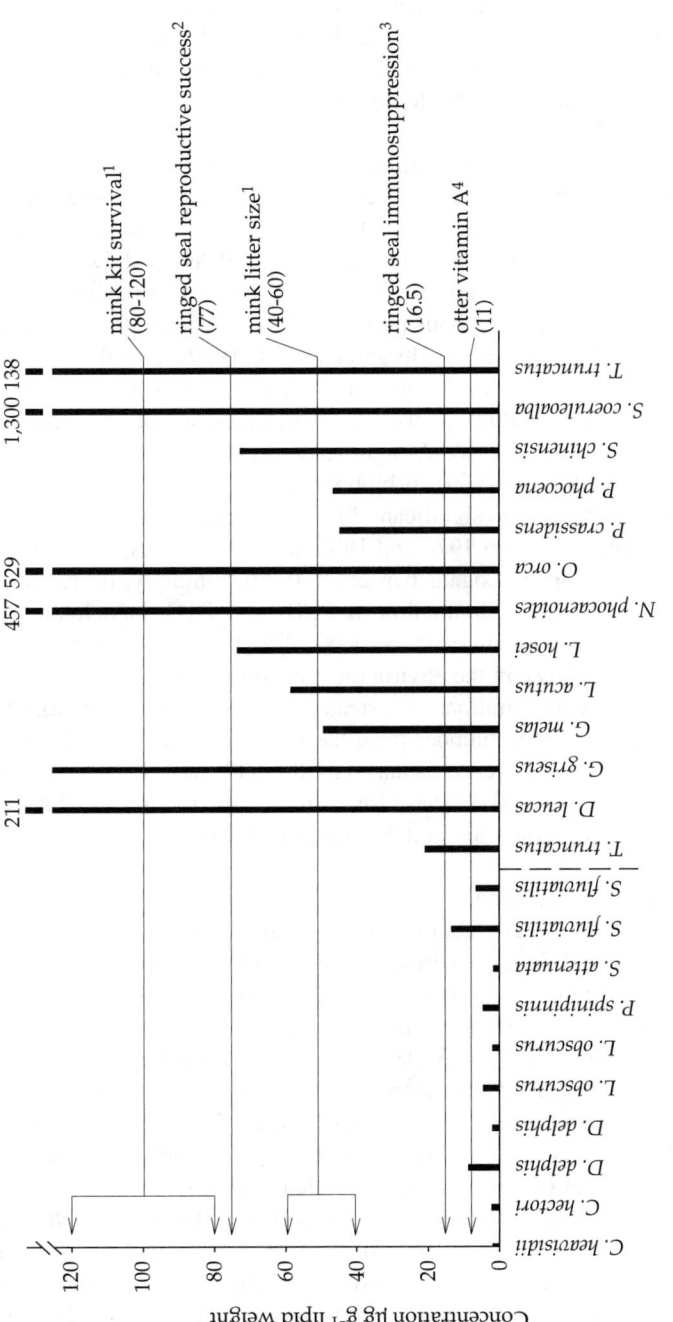

Fig. 2. ΣPCB (μg g^{-1} l.w.) in blubber of male delphinoids from the Southern and Northern Hemispheres compared to thresholds for mammalian effects ([1]Leonards et al. 1995; [2]Helle et al. 1976; [3]de Swart et al. 1994, 1996; [4]Murk et al. 1998). Southern Hemisphere: Cockcroft et al. 1989; Corcuera et al. 1995; da Silva et al. 2003; deKock et al. 1994; Jones et al. 1999; Kemper et al. 1994; Yogui et al. 2003. Northern Hemisphere: Bergrenna et al. 1999; Borrell 1993; Jarman et al. 1996; Kannan et al. 1989; Kuehl et al. 1991; Martineau et al. 1987; Minh et al. 1999, 2000; Prudente et al. 1997.

sphere. Extrapolations from these experimental studies provide an overview of the possible adverse effects of exposure to specific PCB doses. However, because of tissue and species differences, as well as real-life exposure to complex mixtures of contaminants, these comparisons should be used with caution.

Temporal Trends of PHCs in Stranded, Hunted, and By-Catch Animals There are few temporal trends studies of PHCs in cetaceans. The long-term assessment of environmental pollutants in stranded animals enables the analysis of temporal trends of contamination. Decreases in concentrations of PCBs and DDT were observed at the end of the 1980s and the beginning of the 1990s, compared to the late 1970s or the early 1980s, in harbour porpoises from Denmark (Granby and Kinze 1991), Kattegat-Skagerrak Seas (Berggrena et al. 1999), and the Bay of Fundy, Canada (Westgate et al. 1997), and in Dall's porpoises from Japan's coastal waters (Kajiwara et al. 2002), as well as in St. Lawrence belugas (Martineau et al. 1987; Muir et al. 1996b). A significant decline in PCBs was also observed by Stern and Addison (1999) in blubber samples from Eastern Arctic belugas (1982–1997). However, no significant difference was observed in concentrations of PCBs and DDT between 1978 and 1986 in blubber of striped dolphins from the Pacific coast of Japan (Loganathan et al. 1990). Similarly, de Kock et al. (1994) observed no temporal diminution in PCBs and DDT in delphinoids from western South Africa between 1980 and 1987. These results suggest a continuous input of PCBs and DDT in the environment in some regions of the world, large enough to maintain concentrations at a steady state (Muir et al. 1999a). On the other hand, PBDEs in beluga blubber from the Canadian Arctic (1982–1997 and 1989–2001) and the St. Lawrence estuary (1988–1999) have been observed to be increasing (Lebeuf et al. 2004). These trends reflect the production and use of PBDEs in industrialized countries and the impact of long-range transfer of contaminants.

Through metabolic detoxification, p,p'-DDT is converted by dehydrochlorination to p,p'-DDE. A method was elaborated by Aguilar (1984) to determine the chronology of input of DDT in the environment based on the ratio of p,p'-DDE/ΣDDT. A ratio greater than 0.60 signifies that there are no new inputs of DDT in the ecosystem. Greater ratios, indicating limited new inputs of DDT, were observed in common dolphins (range, 0.63–0.91) and harbour porpoises (0.53–0.68) from Irish waters (Smyth et al. 2000), harbour porpoises from the North (0.79) and Baltic Seas (0.75) (Bruhn et al. 1999), and beluga whales from Svalbard, Norway (0.69–0.81) (Andersen et al. 2001). In bottlenose dolphins and common dolphins from the east coast of Australia, ratios varied between 0.6 and 1.0 (Law et al. 2003b). Composition of pesticides in Dall's porpoises from Japan show a decline of major components of DDT and an increase in persistent metabolites (Kajiwara et al. 2002). Likewise, temporal trends in belugas from the Cumberland Sound, Canada, show an increase in ratios, from 0.37 in 1982 to 0.48 in 1997 (Stern 1999). In the St. Lawrence Estuary, greater ratio values were recorded in belugas found stranded between 1987 and 1990 (0.64) (Muir et al. 1996a) than in belugas that died between 1982 and 1985 (0.52) (Martineau et al. 1987), thus indicating a temporal decline in the input of DDT in the estuary ecosystem. In male belugas from West Greenland, p,p'-DDE/ΣDDT ratios (0.56) were similar to those found in male belugas from Hudson Bay (0.57) but higher than ratios from male belugas from the Beaufort Sea (0.49) and Jones Sound (0.55) (Muir et al. 1990; Stern et al. 1994). Lower ratios, indicating new inputs of DDT in the marine environment, were also found in harbour porpoises from Scandinavian waters (0.44) as well as white-sided dolphins (0.57), harbour porpoises (0.55), and long-finned pilot whales (0.57) from the Faroe Islands, North Atlantic (Borrell 1993).

Advantages and Limitations of Studying Stranded, Hunted, and By-Catch Delphinoids The sampling of dead animals creates a unique opportunity to collect morphometric and epidemiological data. However, it is important to keep in mind that samples taken from stranded animals are not entirely representative of the free-ranging population (Aguilar and Borrell 1994b). A comparative study of contaminations between blubber biopsy samples of free-ranging belugas and blubber samples collected from stranded belugas indicated a possible overestimation of contaminants when only samples of dead animals were considered (Hobbs et al. 2003). Stranded animals that have died of natural causes often have died of diseases that may involve alteration of functions of detoxification organs as the liver or kidney, with subsequent secondary infections or starvation (Aguilar and Borrell 1994b). Furthermore, major changes in tissue composition occur rapidly after death and may bias contaminant load (Borrell and Aguilar 1990). Additionally, stranded animal sampling is not homogeneous for all ages and sexes but is usually represented by neonates, near-term females, and older animals (Aguilar and Borrell 1994b). Unrepresentative sampling may also occur when studying hunted or by-catch animals. Hunting regulations, for example, often prohibit the capture of calves, juveniles, or females, and by-catch animals may be mostly represented by inexperienced juveniles or sick individuals.

Nonetheless, studying stranded animals is a practical and relatively inexpensive way to collect valuable data on cetaceans. Efforts to assess the biological effects of PHCs, rather than merely measuring concentrations, have increased over the last decade. More reliable methods are being developed to monitor wild populations adequately. Collection of biological samples from free-living individuals enables the assessment of contaminants and the analysis of biomarkers of exposure and effect, in addition to the recording of morphometric and genealogic data of free-ranging populations. Studying PHCs and their effects in wild marine mammals may give a more realistic and complete view of the impacts of contaminants in wildlife compared to stranded animals. However, these studies are very expensive and require extensive logistics.

B. Free-Ranging Delphinoid Populations

Relatively few attempts have been made to assess PHCs in free-ranging cetaceans. Analyses of blubber biopsies, collected from live-captured animals or by remote sampling using darts, are used as indicators of the total amount of body contaminants (Aguilar and Borrell 1994b; Andersen et al. 2001; Gauthier et al. 1997; Hobbs et al. 2003; Norstrom et al. 1998). These noninvasive methods have been used since the beginning of 1990s in different wild marine mammal populations (e.g., walrus, dolphin, killer whale, and beluga). Skin has also been identified as an indicator tissue to assess PHCs in cetaceans and pinnipeds (Aguilar and Borrell 1994b; Wiig et al. 2000), as well as a reliable tissue to monitor nondestructive biomarkers such as mixed-function oxidase activities (Fossi et al. 1992, 1999). Other biological samples, such as blood, milk, urine, feces, and fur in pinnipeds, may also be used to study the toxicokinetics of contaminants and monitor biomarkers (e.g., immune and endocrine responses in blood) (Fossi et al. 1999). PHCs from blubber biopsies of wild delphinoid populations are presented in Table 8.

Spatial and Temporal Trends of PHCs in Free-Ranging Animals In free-ranging delphinoids, highest loads of PCBs were found in the southern resident (geometric mean of PCBs in males, 146 $\mu g\ g^{-1}$ l.w.; in females, 55 $\mu g\ g^{-1}$ l.w.) and transient killer whales from British Columbia (geometric mean of PCBs in males, 251 $\mu g\ g^{-1}$ l.w.; in females, 59 $\mu g\ g^{-1}$ l.w.) (Ross et al. 2000). These populations are listed among the most contaminated free-ranging marine mammal populations in the world, along with the transient killer whale population from Alaska (mean PCBs, 230 $\mu g\ g^{-1}$ l.w.; mean DDTs, 320 $\mu g\ g^{-1}$ l.w.) (Ylitalo et al. 2001) and the Mediterranean striped dolphin (PCBs, 282 $\mu g\ g^{-1}$ l.w. and 15.5–86 $\mu g\ g^{-1}$ d.w.; DDTs, 15.6–63.6 $\mu g\ g^{-1}$ d.w.) (Aguilar and Borrell 1994a; Marsili and Focardi 1997). Significant levels of PCBs were also detected in blubber of adult male bottlenose dolphins from Texas (91.2 $\mu g\ g^{-1}$ l.w.) and Florida (76.2 $\mu g\ g^{-1}$ l.w.) (Schwacke et al. 2002) as well as South Carolina (50.4 $\mu g\ g^{-1}$ l.w.) and North Carolina (44.2 $\mu g\ g^{-1}$ l.w.) (Hansen et al. 2004). These free-living delphinoid populations were more contaminated than St. Lawrence belugas (mean PCBs in males, 7.6 and 11.8 $\mu g\ g^{-1}$ l.w.) (Hobbs et al. 2003; Letcher et al. 2000b) and the Shannon estuary bottlenose dolphins (mean PCBs in males, 29.5 $\mu g\ g^{-1}$ l.w.) (Berrow et al. 2002). The lowest concentrations of PHCs were found in blubber of live-captured belugas from Svalbard, Norway (mean PCBs and DDT, 5.1 $\mu g\ g^{-1}$ l.w.) (Andersen et al. 2001) and West Hudson Bay (ΣPCBs: 4.3 $\mu g\ g^{-1}$ l.w.) (Letcher et al. 2000b). HCB, CHLs, and HCHs were analyzed in three populations and detected in lowest levels (Andersen et al. 2001; Berrow et al. 2002; Hobbs et al. 2003).

Comparison between killer whale populations showed much higher concentrations of organic pollutants in biopsies from British Columbia and Alaska transient killer whales than residents, which is probably due to the difference in diet (fish-eating versus marine mammal-eating killer whales) (Ross et al. 2000; Ylitalo et al. 2001). The recruitment order is also a factor influencing contamination loads. First-recruit adult male resident Alaskan killer whales, which probably received a higher load of contaminants from their first-time mothers, contained much higher PHC concentrations than those measured in non-first-recruited resi-

Table 8. Mean Concentration and Standard Deviation (SD) ($\mu g\ g^{-1}$ Lipid Weight) of Selected PHCs in Blubber Biopsies of Free-Ranging Delphinoids.

Species and Location	N, Sex	Age or Class	Year	ΣPCBs	ΣDDTs	HCB	ΣCHLs	ΣHCHs	Reference
Delphinus delphis									
Atlantic coast, Spain	22	—	1984	31.1 ± 18.2^{20}	15.4 ± 8.5^4	—	—	—	Borrell et al. 2001
	33m	—	1996	37.9 ± 19^{20}	9.5 ± 4.2^4	—	—	—	
	18f	—	1996	23.9 ± 17.7^{20}	5.1 ± 3.1^4	—	—	—	
Delphinapterus leucas									
W. Hudson Bay, Canada	7m	—	1993–1994	4.3	—	—	—	—	Letcher et al. 2000b***
St. Lawrence estuary	30m	8–22	1994–1996	7.6	—	—	—	—	
	3f	7–15	1994–1996	10.9	—	—	—	—	
Svalbard, Norway	10m	a	1995–1997	5.1 ± 1.9^{27}	5.1 ± 1.1^3	0.5 ± 0.3	2.9 ± 1.2^5	0.2 ± 0.1^3	Andersen et al. 2001
St. Lawrence estuary	34m	8–22	1994–1997	11.8^{104}	9.4^6	0.2	2^7	0.5^3	Hobbs et al. 2003[c]
	10f	7–22	1995–1998	12.2^{104}	3.6^6	0.1	1.4^7	0.3^3	
Orcinus orca									
British Columbia, Canada	Northern resident		1993–1996						Ross et al. 2000[b]
	8m	a		37.4 ± 6.1^{136}	—	—	—	—	
	9f	a		9.3 ± 2.8^{136}	—	—	—	—	
	Southern resident		1993–1996						
	4m	—		146.3 ± 32.7^{136}	—	—	—	—	
	2f	—		55.4 ± 19.3^{136}	—	—	—	—	
	Transient		1993–1996						
	5m	—		251.2 ± 54.7^{136}	—	—	—	—	
	5f	—		58.8 ± 20.6^{136}	—	—	—	—	
Kenai Fjords/Prince William Sound, AK, USA	Resident 29m, 31f, 4u	j,a	1994–1999	14 ± 13	13 ± 14^5	—	—	—	Ylitalo et al. 2001
	Transient 6m, 6f, 1u	j,a	1994–1999	230 ± 130	320 ± 210^5	—	—	—	

Table 8. (Continued).

Species and Location	N, Sex	Age or Class	Year	ΣPCBs	ΣDDTs	HCB	ΣCHLs	ΣHCHs	Reference
Stenella coeruleoalba									
Mediterranean Sea	109	—	1987–1991	282	—	—	—	—	Aguilar and Borrell 1994a[a]
Mediterranean Sea, Italy	89	—	1990–1993	15.5–86[30] Range	15.6–63.5[6] Range	—	—	—	Marsili and Focardi 1996[d]
Tursiops truncatus									
Shannon estuary, Ireland	6m	—	2000	29.5 ± 15.9[7]	20.1 ± 15.5[e]	0.2 ± 0.2	4.2 ± 4.1[2]	0.6 ± 0.6[3]	Berrow et al. 2002
	2f	—	2000	7.1[7]	3.8[e]	0.2	1.4[2]	0.2[3]	
Beaufort, NC, USA	21	j	1995	38.3 ± 14	—	—	—	—	Schwacke et al. 2002
	2m	a		70.3 ± 7.9	—	—	—	—	
	4f	a		4.2 ± 1.2	—	—	—	—	
Matagorda Bay, TX, USA	19	j	1992	86.2 ± 74.4	—	—	—	—	
	7m	a		91.2 ± 53.6	—	—	—	—	
	7f	a		10.3 ± 14.5	—	—	—	—	
Sarasota, FL, USA	19	j	1997–1999	76.8 ± 62.9	—	—	—	—	
	6m	a		76.2 ± 84	—	—	—	—	
	11f	a		5.1 ± 4.3	—	—	—	—	
Beaufort, NC, USA	29	j	1995–2000	27.2[15]	28.9[6]	0.06	3.9[4]	—	Hansen et al. 2004[c]
	5m	a	1995–2000	44.2[15]	51.9[6]	—	7[4]	—	
	6f	a	1995–2000	5.6[15]	4[6]	—	0.6[4]	—	
Charleston, SC, USA	2	j	1999–2001	27.4[27]	17.8[6]	0.05	3.4[4]	—	
	4m	a	1999–2001	50.4[27]	33.1[6]	—	6.6[4]	—	
	5f	a	1999–2001	8[27]	4.5[6]	—	0.7[4]	—	
Indian River Lagoon, FL, USA	9m	—	2002	20.0[27]	12.7[6]	—	2.9[4]	—	
	2f	—	2002	9.3[27]	4.3[6]	—	1[4]	—	

***data include biopsies from live-captured and hunted animals.

Note: Superscript numbers indicate the total number of compounds included in the sum.

m, male; f, female; u, unknown; a, adult; j, juvenile; [a]Medians; [b]Mean and standard error; [c]Geometric means; [d]Data expressed in µg g^{-1} of dry weight; [e]p,p'-DDE only.

dent animals from the same age group (Ylitalo et al. 2001). In addition, several biological factors such as age, sex, and reproductive status also affected the pollutant loads of killer whales from both locations (Ross et al. 2000; Ylitalo et al. 2001).

The assessment of environmental contaminants in blubber samples from free-living cetaceans can also be useful in assessing health risks associated with exposure to PHCs. Recently, Schwacke et al. (2002) developed a novel risk assessment approach integrating PCBs in blubber of free-ranging bottlenose dolphins from the southeast U. S. coast (see Table 8) and surrogate dose–response relationship based on exposure of experimental mink. Their analysis predicts a significant risk of fetal and neonatal mortality for first-time mothers in populations from Matagorda Bay, TX, Sarasota, FL, and Beaufort, NC. This method for risk assessment based on mink laboratory studies may provide hints about the potential effects of PHCs in cetaceans but needs to be cautiously interpreted as mink are some of the most sensitive species to PHCs (Giesy and Kannan 1998). Nonetheless, mean total PCBs found in blubber biopsies of bottlenose dolphins from the American locations, fall within ranges of PCBs found in all other free-ranging delphinoid species reported in this review. These observations reflect the threat of chemical contamination on free-ranging delphinoid populations.

In Vitro Assays Capture–release procedures of wild cetaceans enable the possibility to collect valuable biological samples (e.g., blood, blubber) that may be used in *in vitro* immunological and endocrinological biomarker assays. De Guise et al. (1998) evaluated the effects of *in vitro* exposure of free-ranging arctic beluga whale peripheral blood leukocytes and splenocytes to different persistent PHCs. Reduced proliferation of beluga splenocytes was observed when they were exposed to certain PCB congeners (CB-138, 153, 180, and 169), mixture of congeners, or DDT metabolites (p,p'-DDT and p,p'-DDE). No marked effect on phagocytosis (i.e., the process in the immune response by which undesirable material is ingested by cells, usually neutrophils and monocytes) was observed when peripheral blood leukocytes and splenocytes of beluga were exposed to contaminants (De Guise et al. 1998). A reduced *in vitro* mitogen-induced lymphocyte proliferation response, associated with increasing blood concentrations of PCBs and DDT and its metabolites, was also observed in five captured and released bottlenose dolphins from Sarasota Bay, Florida (Lahvis et al. 1995). Results, similarly to experiments on laboratory mammals and captive seals, suggest an immunosuppression caused by environmental exposure to PHCs.

Advantages and Limitations of Studying Free-Ranging Delphinoids Collecting biological samples from free-ranging marine mammal individuals has many advantages. Biological samples of free-living animals, preferentially samples from animals living in pristine areas, can help us understand the basic biochemistry of marine mammals and the impacts of natural factors, i.e., age, sex, and season, on their physiological systems. Samples can also be used for isotopic studies of diets (Ylitalo et al. 2001). Studying live animals, as opposed to deceased ones, assures more representative samples, and numerous factors such as the number of samples, the dates of sampling, and geographic locations can be controlled by the investigators (Hobbs et al. 2003). Samples may also be used for genomic, immunological or endocrinological assays which can help determine the impacts of these contaminants on marine mammals (Erickson et al. 1995; Fossi et al. 1999; Gauthier et al. 1999; Mos and Ross 2002), as well as help elaborate cause–effect relationships (Muir et al. 1999a).

Biological samples used during contaminant monitoring should be carefully chosen according to the purpose of the research. In a study by Lydersen et al. (2002), blood has been found to be a poor substrate for monitoring time trends of organochlorine contaminants in phocid seal populations because of the extreme variability in contaminant concentrations in response to change in condition (i.e., during breeding and molting, phocid seals experience extended periods of fasting that can result in a body mass depletion of as much as 40%). Nonetheless, remote biopsy sampling enables the unique collection of valuable data on threatened or endangered species without causing disturbance or stress to the population (Hobbs et al. 2003).

However, the biopsy method also has its limitations. Distribution and composition of lipids in the blubber of cetaceans is not homogeneous (Addison 1989; Aguilar 1987; Aguilar and Borrell 1994b; Hobbs et al. 2003), and variations can occur both along and through the blubber layer (Kawai et al. 1988). Therefore, to detect levels of contaminants that reflect the real body load, samples from a specific body location containing all strata of the blubber layer need to be collected and lipid content and composition of the sample considered (Aguilar and Borrell 1994b). Another concern about biopsying wild animals is the difficulty of estimating biological parameters (i.e., age, sex, reproductive status, and recruitment order) of individuals, which is important in quantifying environmental trends of PHCs. This consideration leads to the importance of long-term monitoring studies of marine mammal populations. For example, the Sarasota Bay bottlenose dolphin community, located on the west central coast of Florida, has been the subject of a long-term study since 1970, and data on four generations have been collected (Wells 1999). Other delphinoid populations such as killer whales from Alaska (Matkin et al. 1999) and beluga whales from the St. Lawrence estuary have also been the subject of long-term studies (Béland et al. 1993). Biopsy samples from those free-ranging and known individuals represent unique opportunities for research.

V. Conclusion

Mounting evidence suggests that chemical environmental contaminants, such as PHCs, are factors that contribute to the overall degradation of several marine mammal populations. Although there is evidence of decline of PCBs and DDT in the marine ecosystem, there are increasing concentrations of PBDEs and reports of elevated levels of PFAs, which emphasizes the importance of monitoring the health of marine mammals and their environment. Our current and future knowledge of the impacts of environmental contaminants on marine mammals is the result of an ensemble of methods. Monitoring of stranded individuals enables the collection of important morphometric and epidemiological data. The use of noninvasive methods such as blood and skin/blubber collection, permits a more realistic estimation of the contaminant loads of wild animals and helps monitor environmental changes. Long-term monitoring studies of marine mammal populations also give valuable temporal trend data. The use of biomarkers, surrogate species, and laboratory mammals, as well as *in vitro* studies, help us to understand mechanisms of action and impacts of pollutants on the metabolic, immunological, endocrine, and reproductive systems of marine mammals as well as to elaborate cause–effect links.

Summary

This chapter reviews the global distribution, biotransformation, accumulation patterns, and mechanisms of action and the potential impacts of persistent organohalogen contaminants (PHCs) on physiological systems of cetaceans with emphasis on delphinoids. Methods used to study PHCs in stranded and free-living cetaceans are discussed, and concentrations of PHCs of stranded, hunted, bycatch, and free-ranging delphinoids are summarized. Overall, the highest concentrations of PHC contamination were found in delphinoids from industrialized areas of the Northern Hemisphere compared to the Southern Hemisphere. Nonetheless, PHCs are also found in marine mammal tissues from the Southern Hemisphere and in remote regions such as the Arctic, reflecting the global distribution and contamination of PHCs in the marine ecosystem.

Acknowledgments

The authors thank the Natural Sciences and Engineering Research Council of Canada (NSERC) and the Fonds de recherche sur la nature et les technologies (Québec) for funding provided to M.H. as well as R. Letcher for his review of the manuscript. Merci.

References

Addison RF, Brodie PF (1977) Organochlorine residues in maternal blubber, milk, and pup blubber from grey seals (*Halichoerus grypus*) from Sable Island, Nova Scotia. J Fish Res Board Can 34:937–941.

Addison RF (1989) Organochlorines and marine mammal reproduction. Can J Fish Aquat Sci 46:360–368.

Aguilar A (1984) Relationship of DDE/ΣDDT in marine mammals to the chronology of DDT input into the ecosystem. Can J Fish Aquat Sci 41:840–844.

Aguilar A (1987) Using organochlorine pollutants to discriminate marine mammal populations: a review and critique of the methods. Mar Mamm Sci 3:242–262.

Aguilar A, Borrell A (1994a) Abnormally high polychlorinated biphenyl levels in striped dolphins (*Stenella coeruleoalba*) affected by the 1990–1992 Mediteranean epizootic. Sci Total Environ 154:237–247.

Aguilar A, Borrell A (1994b) Assessment of organochlorine pollutants in cetaceans by means of skin and hypodermic biopsies. In: Fossi MC, Leonzio C (eds) Nondestructive Biomarkers in Vertebrates. CRC Press, Boca Raton, FL, pp 245–267.

Aguilar A, Raga, JA (1993) The striped dolphin epizootic in the Mediterranean Sea. Ambio 22:524–528.

Aguilar A, Borrell A, Reijnders PJH (2002) Geographical and temporal variation in levels of organochlorine contaminants in marine mammals. Mar Environ Res 53:425–452.

Alaee M, Sergeant DB, Muir DCG, Whittle DM, Solomon KR (1999) Distribution of polybrominated diphenyl ethers in the Canadian environment. Organohalogen Compd 40:347–350.

Andersen G, Kovacs KM, Lydersen C, Skaare JU, Gjertz I, Jenssen BM (2001) Concentrations and patterns of organochlorine contaminations in white whales (*Delphinapterus leucas*) from Svalbard, Norway. Sci Total Environ 264:267–281.

Bäcklin B-M, Bergman A (1992) Morphological aspects on the reproductive organs in female mink (*Mustela vison*) exposed to polychlorinated biphenyls and fractions thereof. Ambio 10:596–601.

Bailey RE (2001) Global hexachlorobenzene emissions. Chemosphere 43:167–182.

Banerjee BD (1987) Effects of sub-chronic DDT exposure on humoral and cell-mediated immune responses in albino rats. Bull Environ Contam Toxicol 39:827–834.

Banerjee BD, Ramachandran M, Hussain QZ (1986) Sub-chronic effect of DDT on humoral immune response in mice. Bull Environ Contam Toxicol 37:433–440.

Beck H, Breuer EM, Drob A, Mathar W (1990) Residues of PCDDs, PCDFs, PCBs, and other organochlorine compounds in harbour seals and harbour porpoises. Chemosphere 20:1027–1034.

Becker PR, Mackey EA, Demiralp R, Schantz MM, Koster BJ, Wise SA (1997) Concentrations of chlorinated hydrocarbons and trace element in marine mammal tissues archived in the U.S. national biomonitoring specimen bank. Chemosphere 34:2067–2098.

Béland P, De Guise S, Girard C, Lagacé A, Martineau D, Michaud R, Muir DCG, Norstrom RJ, Pelletier E, Ray S, Shugart LR (1993) Toxic compounds and health and reproductive effects in St. Lawrence beluga whales. J Great Lakes Res 19:766–775.

Bennett E, Ross PS, Letcher RJ (2002) Polyhalogenated phenolic contaminants in Pacific killer whale (*Orcinus orca*). Organohalogen Compd 58:81–84.

Berggrena P, Ishaq R, Zebühr Y, Näf C, Bandh C, Broman D (1999) Patterns and levels of organochlorines (DDTs, PCBs, non-*ortho* PCBs and PCDD/Fs) in male harbour porpoises (*Phocoena phocoena*) from the Baltic Sea, the Kattegat-Skagerrak Seas and the west coast of Norway. Mar Pollut Bull 38:1070–1084.

Bergman Å, Norstrom RJ, Haraguchi K, Kuroki H, Béland P (1994) PCB and DDE methyl sulfones in mammals from Canada and Sweden. Environ Toxicol Chem 13:121–128.

Berrow SD, Mchugh B, Glynn D, Mcgovern E, Parsons KM, Baird RW, Hooker SK (2002) Organochlorine concentrations in resident bottlenose dolphins (*Tursiops truncatus*) in the Shannon estuary, Ireland. Mar Pollut Bull 44:1296–1313.

Bleavins MR, Aulerich RJ, Ringer RK (1980) Polychlorinated biphenyls (Aroclors 1016 and 1242): effects on survival and reproduction in mink and ferrets. Arch Environ Contam Toxicol 9:627–635.

Boon JP, van der Meer J, Allchin CR, Law RJ, Klungsøyr J, Leonards PEG, Spliid H, Storr-Hansen E, McKenzie C, Wells DE (1997) Concentration-dependent changes of PCB patterns in fish-eating mammals: structural evidence for induction of cytochrome P450. Arch Environ Contam Toxicol 33:298–311.

Boon JP, Lewis WE, Goksøyr A (2001) Immunochemical and catalytic characterization of hepatic microsomal cytochrome P450 in the sperm whale (*Physeter macrocephalus*). Aquat Toxicol 52:297–309.

Boon JP, Lewis WE, Tjoen-A-Choy MR, Allchin CR, Law RJ, de Boer J, Ten Hallers-Tjabbes CC, Zegers BN (2002) Levels of polybrominated diphenyl ether (PBDE) flame retardants in animals representing different trophic levels of the North Sea food web. Environ Sci Technol 36: 4025–4032.

Borrell A (1993) PCB and DDTs in blubber of cetaceans from the northeastern North Atlantic. Mar Pollut Bull 26:146–151.

Borrell A, Aguilar A (1990) Loss of organochlorine compounds in the tissues of a decomposing stranded dolphin. Bull Environ Contam Toxicol 45:46–53.

Borrell A, Bloch D, Desportes G (1995) Age trends and reproductive transfer of organochlorine compounds in long-finned pilot whales from the Faroe Islands. Environ Pollut 88:283–292.

Borrell A, Aguilar A, Corsolini S, Focardi S (1996) Evaluation of toxicity and sex-related variation of PCB levels in Mediterranean striped dolphins affected by an epizootic. Chemosphere 32:2359–2369.

Borrell A, Cantos G, Pastor T, Aguilar A (2001) Organochlorine compounds in common dolphins (*Delphinus delphis*) from the Atlantic and Mediterranean waters of Spain. Environ Pollut 114:265–274.

Brandt I, Jönsson C-J, Lund B-O (1992) Comparative studies on adrenocorticolytic DDT metabolites. Ambio 21:602–605.

Breivik K, Sweetman A, Pacyna JM, Jones KC (2002a) Towards a global historical emission inventory for selected PCB congeners: a mass balance approach. 2. Emissions. Sci Total Environ 290:199–224.

Breivik K, Sweetman A, Pacyna JM, Jones KC (2002b) Towards a global historical emission inventory for selected PCB congeners: a mass balance approach. 1. Global production and consumption. Sci Total Environ 290:181–198.

Brouwer A, Reijnders PJH, Koeman JH (1989) Polychlorinated biphenyl (PCB)-contaminated fish induces vitamin A and thyroid hormone deficiency in the common seal (*Phoca vitulina*). Aquat Toxicol 15:99–106.

Bruhn R, Kannan N, Petrick G, Schulz-Bull DE, Duinker JC (1999) Persistent chlorinated organic contaminants in harbour porpoises from the North Sea, the Baltic sea and Arctic waters. Sci Total Environ 237/238:351–361.

Cockcroft VG, de Kock AC, Lord DA, Ross GJB (1989) Organochlorines in bottlenose dolphins *Tursiops truncatus* from the east coast of South Africa. S Afr J Mar Sci 8: 207–217.

Cockcroft VG, de Kock AC, Ross GJB, Lord DA (1990) Organochlorines in common dolphins caught in shark nets during the Natal 'sardine run.' S Afr J Zool 25:144–148.

Connell DW, Miller GJ, Mortimer MR, Shaw GR, Anderson SM (1999) Persistent lipophilic contaminants and other chemical residues in Southern Hemisphere. Crit Rev Envir Sci Technol 29:47–82.

Corcuera J, Monzón F, Aguilar A, Borrell A, Raga JA (1995) Life history data, organochlorine pollutants and parasites from eight Burmeister's porpoises *Phocoena spinipinnis*, caught in northern Argentine waters. Rep Int Whal Comm Special Issue 16: 365–372.

Corsolini S, Focardi S, Kannan K, Tanabe S, Borrell A, Tatsukawa R (1995) Congener profile and toxicity assessment of polychlorinated biphenyls in dolphins, sharks and tuna collected from Italian coastal waters. Mar Environ Res 40:33–53.

Covaci A, van de Vijver K, DeCoen W, Das K, Bouquegneau JM, Blust R, Schepens P (2002) Determination of organohalogenated contaminants in liver of harbour porpoises (*Phocoena phocoena*) stranded on the Belgian North Sea coast. Mar Pollut Bull 44:1152–1169.

Dam M, Boch D (2000) Screening of mercury and persistent organochlorine pollutants in long-finned pilot whale (*Globicephala melas*) in the Faroe Islands. Mar Pollut Bull 40:1090–1099.

Darnerud PO, Eriksen GS, Johannesson T, Larsen PB, Viluksela M (2001) Polybrominated diphenyl ethers:occurrence, dietary exposure, and toxicology. Environ Health Perspect 109:49–68.

da Silva AMF, Lemes VRR, Barretto HHC, Oliveira ES, de Alleluia IB, Paumgartten FJR (2003) Polychlorinated biphenyls and organochlorine pesticides in edible fish species and dolphins from Guanabara Bay, Rio de Janeiro, Brazil. Bull Environ Contam Toxicol 70:1151–1157.

de Boer J, Wester PG, Klamer HJC, Lewis WE, Boon JP (1998) Do flame retardants threaten ocean life? Nature (Lond) 394:28–29.

De Guise S, Lagacé A, Béland P (1994) Tumors in St. Lawrence beluga whales (*Delphinapterus leucas*). Vet Pathol 31:444–449.

De Guise S, Martineau D, Béland P, Fournier M (1998) Effects of *in vitro* exposure of beluga whale leukocytes to selected organochlorines. J Toxicol Environ Health Part A 55:479–493.

de Kock AC, Best PB, Cockcroft V, Bosma C (1994) Persistent organochlorine residues in small cetaceans from the east and west coasts of southern Africa. Sci Total Environ 154:153–162.

DeLong RL, Gilmartin WG, Simpson JG (1970) Premature births in California sea lions: association with high organochlorine pollutant residue levels. Science 181:1168–1169.

de March BGE, de Wit CA, Muir DCG, Braune BM, Gregor DJ, Norstrom RJ, Olsson M, Skaare JU, Strange K (1998) Persistent organic pollutants. In: de March BGE, de Wit CA, Muir DCG (eds) Arctic Monitoring and Assessment Programme (AMAP). Assessment Report: Arctic Pollution Issues. AMAP, Oslo, Norway, pp 183–371.

de Swart RL, Ross PS, Vedder LJ, Timmerman HH, Heisterkamp S, Van Loveren SH, Vos JG, Reijnders PJH, Osterhaus ADME (1994) Impairment of immune function in harbor seals (*Phoca vitulina*) feeding on fish from polluted waters. Ambio 23: 155–159.

de Swart RL, Ross PS, Vedder LJ, Boink FBTJ, Reijnders PJH, Mulder PGH, Osterhaus ADME (1995a) Haematology and clinical chemistry values of harbour seals (*Phoca vitulina*) fed environmentally contaminated herring remain within normal ranges. Can J Zool 73:2035–2043.

de Swart R, Ross PS, Timmerman HH, Vos HW, Reijnders PJH, Vos JG, Osterhaus ADME (1995b) Impaired cellular immune response in harbour seals (*Phoca vitulina*) feeding on environmentally contaminated herring. Clin Exp Immunol 101:480–486.

de Swart RL, Ross PS, Vos JG, Osterhaus ADME (1996) Impaired immunity in harbour seals (*Phoca vitulina*) exposed to bioaccumulated environmental contaminants: review of a long-term study. Environ Health Perspect 104:823–828.

de Wit CA, Fisk AT, Hobbs KE, Muir DCG, Gabrielsen GW, Kallenborn R, Krahn MM, Norstrom RJ, Skaare JU (2004) Persistent organic pollutants in the Arctic. In: de Wit CA, Fisk AT, Hobbs KE, Muir DCG (eds) Arctic Monitoring and Assessment Programme (AMAP), Oslo, Norway, xvi + 310 pp.

Elliot JE, Norstrom RJ, Lorenzen A, Hart LE, Philibert H, Kennedy SW, Stegeman JJ, Bellward GD, Cheng KM (1996) Biological effects of polychlorinated dibenzo-*p*-dioxins, dibenzofurans, and biphenyls in bald eagle (*Haliaeetus leucocephalus*) chicks. Environ Toxicol Chem 15:782–793.

Erickson KL, DiMolfetto-Landon L, Wells RS, Reidarson T, Stott JL, Ferrick DA (1995) Development of an interleukin-2 receptor expression assay and its use in evaluation of cellular immune responses in bottlenose dolphin (*Tursiops truncatus*). J Wildl Dis 31:142–149.

Falandysz J, Yamashita N, Tanabe S, Tatsukawa R, Rucińska L, Skóra K (1994) Congener-specific data on polychlorinated biphenyls in tissues of common porpoise from Puck Bay, Baltic Sea. Arch Environ Contam Toxicol 26:267–272.

Ford CA, Muir DCG, Norstrom RJ, Simon M, Mulvihill MJ (1993) Development of a semi-automated method for non-ortho PCBs: application to Canadian arctic marine mammal tissues. Chemosphere 26:1981–1991.

Fossi MC, Marsili L, Leonzio C, Di Sciara GN, Zanardelli M, Focardi S (1992) The use of non-destructive biomarker in Mediterranean cetaceans: preliminary data on MFO activity in skin biopsy. Mar Pollut Bull 24:459–461.

Fossi MC, Casini S, Marsili L (1999) Nondestructive biomarkers of exposure to endocrine disrupting chemicals in endangered species of wildlife. Chemosphere 39:1273–1285.

Friend M, Trainer DO (1970) Polychlorinated biphenyl: interaction with duck hepatitis virus. Science 170:1314–1316.

Gauthier JM, Metcalfe CD, Sears R (1997) Validation of the blubber biopsy technique for monitoring of organochlorine contaminants in balaenopterid whales. Mar Environ Res 43:157–179.

Gauthier JM, Pelletier E, Brochu C, Moore S, Metcalfe CD, Béland P (1998) Environmental contaminants in tissues of a neonate St Lawrence beluga whale (*Delphinapterus leucas*). Mar Pollut Bull 36:102–108.

Gauthier JM, Dubeau H, Rassart É, Jarman WM, Wells RS (1999) Biomarkers of DNA damage in marine mammals. Mutat Res 444:427–439.

Giesy JP, Kannan K (1998) Dioxin-like and non-dioxin like toxic effects of polychlorinated biphenyls (PCBs): implications for risk assessment. Crit Rev Toxicol 28:511–569.

Giesy JP, Kannan K (2001) Global distribution of perfluorooctane sulfonate in wildlife. Environ Sci Technol 35:1339–1342.

Giesy JP, Kannan K (2002) Perfluorinated surfactants in the environment. Environ Sci Technol 36:146A–152A.

Goksøyr A (1995) Cytochrome P450 in marine mammals: isozyme forms, catalytic functions, and physiological regulations. In: Blix AS, Walløe L, Ulltang Ø (eds) Whales, Seals, Fish and Man. Proceedings of the International Symposium on the Biology of Marine Mammals in the North East Atlantic, Tromsø, Norway, 29 November–1 December 1994. Elsevier, Amsterdam, pp 629–639.

Grachev MA, Kumarev VP, Mamaev LV, Zorin VL, Baranova LV, Denikina NN, Belikov SI, Petrov EA, Kolesnik VS, Kolesnik RS, Dorofeev VM, Beim AM, Kudelin VN, Nagieva FG, Sidorov VN (1989) Distemper virus in Baikal seals. Nature (Lond) 338:209.

Granby K, Kinze CC (1991) Organochlorines in Danish and west Greenland harbour porpoises. Mar Pollut Bull 22:458–462.

Grasman KA, Fox GA, Scanlon PF, Ludwig JP (1996) Organochlorine-associated immunosuppression in prefledgling Caspian terns and herring gulls from the Great Lakes: an ecoepidemiological study. Environ Health Perspect 104(Suppl 4):829–842.

Gulland FMD, Trupkiewicz JG, Spraker TR, Lowenstine LJ (1996) Metastatic carcinoma of probable transitional cell origin in 66 free-ranging California sea lions (*Zalophus californianus*), 1979 to 1994. J Wildl Dis 32:250–258.

Haglund PS, Zook DR, Buser H-R, Hu J (1997) Identification and quantification of polybrominated diphenyl ethers and methoxy-polybrominated diphenyl ethers in Baltic biota. Environ Sci Technol 31:3281–3287.

Hagmar L, Rylander L, Dyremark E, Klasson-Wehler E, Erfurth EM (2001) Plasma concentrations of persistent organochlorines in relation to thyrotropin and thyroid hormone levels in women. Int Arch Occup Environ Health 74:184–188.

Hakansson H, Manzoor E, Ahlborg UG (1992) Effects of technical PCB preparations and fractions thereof on vitamin A levels in the mink (*Mustela vison*). Ambio 21: 588–590.

Hakk H, Letcher RJ (2003) Metabolism in the toxicokinetics and fate of brominated flame retardants: A review. Environ Int 29:801–828.

Hansen LJ, Schwacke LH, Mitchum GB, Hohn AA, Wells RS, Zohman ES, Fair PA (2004) Geographic variation in polychlorinated biphenyl and organochlorine pesticide concentrations in the blubber of bottlenose dolphins from the US Atlantic coast. Sci Total Environ 319:147–172.

Hallgren S, Sinjari T, Håkansson H, Darnerud PO (2001) Effects of polybrominated diphenyl ethers (PBDEs) and polychlorinated biphenyls (PCBs) on thyroid hormone and vitamin A levels in rats and mice. Arch Toxicol 75:200–208.

Harner T, Jantunen LMM, Bidleman TF, Barrie LA, Kylin H, Strachan WMJ, Macdonald RW (2000) Microbial degradation is a key elimination pathway of hexachlorocyclohexanes from the Arctic Ocean. Geophys Res Lett 27:1155–1159.

Hayteas DL, Duffield DA (1997) The determination by HPLC of PCB and *p,p'*-DDE residues in marine mammals stranded on the Oregon coast, 1991–1995. Mar Pollut Bull 34:844–848.

Hayteas DL, Duffield DA (2000) High levels of PBC and *p,p'*-DDE found in the blubber of killer whales (*Orcinus orca*). Mar Pollut Bull 40:558–561.

Heaton SN, Bursian SJ, Giesy JP, Tillitt DE, Render JA, Jones PD, Verbrugge DA, Kubiak TJ, Aulerich RJ (1995) Dietary exposure of mink to carp from Saginaw Bay, Michigan. 1. Effects on reproduction and survival, and the risks to wild mink populations. Arch Environ Contam Toxicol 28:334–343.

Hebert CE, Keenleyside KA (1995) To normalize or not to normalize? Fat is the question. Environ Toxicol Chem 14:801–807.

Heide-Jørgensen M-P, Härkönen T, Dietz R, Thompson PM (1992) Retrospective of the 1998 European seal epizootic. Dis Aquat Organisms 13:37–62.
Helle E, Olsson M, Jensen, S (1976a) DDT and PCB levels and reproduction in ringed seal from the Bothnian Bay. Ambio 5:188–189.
Helle E, Olsson M, Jensen S (1976b) PCB levels correlated with pathological changes in seal uteri. Ambio 5:261–263.
Helm PA, Bidleman TF, Stern GA, Koczanski K (2002) Polychlorinated naphthalenes and coplanar polychlorinated biphenyls in beluga whale (*Delphinapterus leucas*) and ringed seal (*Phoca hispida*) from then eastern Canadian Arctic. Environ Pollut 119: 69–78.
Hobbs KE, Muir DCG, Michaud R, Béland P, Letcher RJ, Norstrom RJ (2003) PCBs and organochlorine pesticides in blubber biopsies from free-ranging St. Lawrence river estuary beluga whales (*Delphinapterus leucas*), 1994–1998. Environ Pollut 122: 291–302.
Hoekstra PF, O'Hara TM, Teixeira C, Backus S, Fisk AT, Muir DCG (2002) Spatial trends and bioaccumulation of organochlorine pollutants in marine zooplankton from the Alaskan and Canadian Arctic. Environ Toxicol Chem 21:575–583.
Hoekstra PF, O'Hara TM, Fisk AT, Borgå K, Solomon KR, Muir DCG (2003) Trophic transfer of persistent organochlorine contaminants (OCs) within an Arctic marine food web from the southern Beaufort-Chukchi Seas. Environ Pollut 124:509–522.
Honkakoski P, Negishi M (2000) Regulation of cytochrome P450 (*CYP*) genes by nuclear receptors. Biochem J 347:321–337.
Ikonomou MG, Fischer M, He T, Addison RF, Smith T (2000) Congener patterns, spatial and temporal trends of polybrominated diphenyl ethers in biota samples from the Canadian west coast and the Northwest Territories. Organohalogen Compd 47:77–80.
Ikonomou MG, Rayne S, Addison RF (2002) Exponential increases of the brominated flame retardants, polybrominated diphenyl ethers, in the Canadian Arctic from 1981 to 2000. Environ Sci Technol 36:1886–1892.
Iwata H, Tanabe S, Sakai N, Tatsukawa R (1993) Distribution of persistent organochlorines in the oceanic air and surface seawater and the role of ocean on their global transport and fate. Environ Sci Technol 27:1080–1098.
Jarman WM, Simon M, Norstrom RJ, Burns SA, Bacon CA, Simonelt BRT, Risebrough RW (1992) Global distribution of tris(4-chlorophenyl)methanol in high trophic level birds and mammals. Environ Sci Technol 26:1770–1774.
Jarman WM, Norstrom RJ, Muir DCG, Rosenberg B, Simon M, Baird RW (1996) Levels of organochlorine compounds, including PCDDS and PCDFS, in the blubber of cetaceans from the west coast of North America. Mar Pollut Bull 32:426–436.
Jensen BA, Hahn ME (2001) cDNA cloning and characterization of a high affinity aryl hydrocarbon receptor in a cetacean, the beluga, *Delphinapterus leucas*. Toxicol Sci 64:41–56.
Jepson PD, Bennett PM, Allchin CR, Law RJ, Kuiken T, Baker JR, Rogan E, Kirkwood JK (1999) Investigating potential associations between chronic exposure to polychlorinated biphenyls and infectious disease mortality in harbour porpoises from England and Wales. Sci Total Environ 243/244:339–348.
Jones PD, Hannah DJ, Buckland SJ, van Maanen T, Leathem SV, Dawson S, Slooten E, van Helden A, Donoghue M (1999) Polychlorinated dibenzo-*p*-dioxins, dibenzofurans and polychlorinated biphenyls in New Zealand cetaceans. J Cetacean Res Manage Special Issue 1:157–167.
Jönsson C-J, Lund B-O, Bergman A, Brandt I (1992) Adrenocortical toxicity of 3-meth-

ylsulphonyl-DDE. 3: Studies in fetal and suckling mice. Reprod Toxicol 6:233–240.

Jönsson C-J, Lund B-O, Brandt I (1993) Adrenocorticolytic DDT-metabolites: studies in mink, *Mustela vison* and otter, *Lutra lutra.* Ecotoxicology 2:41–53.

Kajiwara N, Watanabe M, Tanabe S, Nakamatsu K, Amano M, Miyazaki N (2002) Specific accumulation and temporal trends of organochlorine contaminants in Dall's porpoises (*Phocoenoides dalli*) from Japanese coastal waters. Mar Pollut Bull 44:1089–1099.

Kannan N, Tanabe S, Ono M, Tatsukawa R (1989) Critical evaluation of polychlorinated biphenyl toxicity in terrestrial and marine mammals: increasing impact of non-*ortho* and mono-*ortho* coplanar polychlorinated biphenyls from land and ocean. Arch Environ Contam Toxicol 18:850–857.

Kannan K, Tanabe S, Borrell A, Aguilar A, Focardi S, Tatsukawa R (1993a) Isomer-specific analysis and toxic evaluation of polychlorinated biphenyls in striped dolphins affected by an epizootic in the western Mediterranean Sea. Arch Environ Contam Toxicol 25:227–233.

Kannan K, Falandysz J, Tanabe S, Tatsukawa R (1993b) Persistent organochlorines in harbour porpoises from the Puck Bay, Poland. Mar Pollut Bull 26:162–165.

Kannan K, Koistinen J, Beckmen K, Evans T, Gorzelany JF, Hansen KJ, Jones PD, Helle E, Nyman M, Giesy JP (2001) Accumulation of perfluooctane sulfonate in marine mammals. Environ Sci Technol 35:1593–1598.

Kannan K, Corsolini S, Falandysz J, Oehme G, Focardi S, Giesy JP (2002) Perfluorooctanesulfonate and related fluorinated hydrocarbons in marine mammals, fishes, and birds from coasts of the Baltic and the Mediterranean Seas. Environ Sci Technol 36:3210–3216.

Karlson K, Ishaq R, Becker G, Berggren P, Broman D, Colmsjö A (2000) PCBs, DDTs and methyl sulphone metabolites in various tissues of harbour porpoises from Swedish waters. Environ Pollut 110:29–46.

Kawai S, Fukushima M, Miyazaki N, Tatsukawa R (1988) Relationship between lipid composition and organochlorine levels in the tissues of striped dolphin. Mar Pollut Bull 19:129–133.

Kawano M, Inoue T, Wada T, Hidaka H, Tatsukawa R (1988) Bioconcentration and residue patterns of chlordane compounds in marine animals: invertebrates, fish, mammals, and seabirds. Environ Sci Technol 22:792–797.

Kemper C, Gibbs P, Obendorf D, Marvanek S, Lenghaus C (1994) A review of heavy metal and organochlorine levels in marine mammals in Australia. Sci Total Environ 154:129–139.

Kennedy SW, Kuiken T, Jepson PD, Deaville R, Forsyth M, Barrett T, van de Bildt MWG, Osterhaus ADME, Eybatov T, Duck C, Kydyrmanov A, Mitrofanov I, Wilson S (2000) Mass die-off of Caspian seals caused by canine distemper virus. Emerg Infect Dis 6:637–639.

Key BD, Howell RD, Criddle CS (1997) Fluorinated organics in the biosphere. Environ Sci Technol 31:2445–2454.

Kleivane L, Skaare JU, Bjørge A, de Ruiter E, Reijnders PJH (1995) Organochlorine pesticide residue and PCBs in harbour porpoise (*Phocoena phocoena*) incidentally caught in Scandinavian waters. Environ Pollut 89:137–146.

Krahn MM, Burrows DG, Stein JE, Becker PR, Schantz MM, Muir DCG, O'Hara TM, Rowles T (1999) White whales (*Delphinapterus leucas*) from three Alaskan stocks: concentrations and patterns of persistent organochlorine contaminants in blubber. J Cetacean Res Manage 1:239–249.

Kuehl DW, Haebler R (1995) Organochlorine, organobromine, metal and selenium residues in bottlenose dolphins (*Tursiops truncatus*) collected during an unusual mortality event in the Gulf of Mexico, 1990. Arch Environ Contam Toxicol 28:494–499.

Kuehl DW, Haebler R, Potter C (1991) Chemical residues in dolphins from the U.S. Atlantic coast including Atlantic bottlenose obtained during the 1987/88 mass mortality. Chemosphere 22:1071–1084.

Kuiken T, Bennett PM, Allchin CR, Kirkwood JK, Baker JR, Lockyer CH, Walton MJ, Sheldrick MC (1994) PCBs, cause of death and body condition in harbour porpoises (*Phocoena phocoena*) from British waters. Aquat Toxicol 28:13–28.

Lahvis GP, Wells RS, Kuehl DK, Stewart JL, Rhinehart HL, Via CS (1995) Decreased lymphocyte responses in free-ranging bottlenose dolphins (*Tursiops truncatus*) are associated with increased concentrations of PCBs and DDT in peripheral blood. Environ Health Perspect 103(suppl 4):67–72.

Lair S, Béland P, De Guise S, Martineau D (1997) Adrenal hyperplastic and degenerative changes in beluga whales. J Wildl Dis 33:430–437.

Law RJ, Allchin CR, Morris RJ (1995) Uptake of organochlorines (chlorobiphenyls, dieldrin; total PCBs & DDT) in bottlenose dolphin (*Tursiops truncatus*) from Cardigan Bay, West Wales. Chemosphere 30:547–560.

Law RJ, Allchin CR, Jones BR, Jepson PD, Baker JR, Spurrier CJH (1997) Metals and organochlorines in tissues of a Blainville's beaked whale (*Mesoplodon densirostris*) and a killer whale (*Orcinus orca*) stranded in the United Kingdom. Mar Pollut Bull 34:208–212.

Law RJ, Allchin CR, Bennett ME, Morris S, Rogan E (2002) Polybrominated diphemyl ethers in two species of marine top predators from England and Wales. Chemosphere 46:673–681.

Law RJ, Alaee M, Allchin CR, Boon JP, Lebeuf M, Lepom P, Stern GA (2003a) Levels and trends of polybrominated diphenylethers and other brominated flame retardants in wildlife. Environ Int 29:757–770.

Law RJ, Morris RJ, Allchin CR, Jones BR, Nicholson MD (2003b) Metals and organochlorines in small cetaceans stranded on the east coast of Australia. Mar Pollut Bull 46:1206–1211.

Lebeuf M, Bernt KE, Trottier S, Noël M, Hammill MO, Measures L (2001) *Tris* (4-chlorophenyl) methane and *tris* (4-chlorophenyl) methanol in marine mammals from the Estuary and Gulf of St. Lawrence. Environ Pollut 11:29–43.

Lebeuf M, Gouteux B, Measures L, Trottier S (2004) Levels and temporal trends (1988–1999) of polybrominated diphenyl ethers in beluga whales (*Delphinapterus leucas*) from the St. Lawrence estuary. Canada. Environ Sci Tech 38:2971–2977.

Leonards PEG, de Vries TH, Minnaard W, Stuijfzand S, de Voogt P, Cofino WP, van Straalen NM, van Hattum B (1995) Assessment of experimental data on PCB-induced reproduction inhibition in mink, based on an isomer- and congener-specific approach using 2,3,7,8-tetrachlorodibenzo-*p*-dioxin toxic equivalency. Environ Toxicol Chem 14:639–652.

Letcher RJ, Norstrom RJ, Muir DCG (1998) Biotransformation versus bioaccumulation: sources of methyl sulfone PCB and 4,4′-DDE metabolites in the polar bear food chain. Environ Sci Technol 32:1656–1661.

Letcher RJ, Klasson-Wehler E, Bergman Å (2000a) Methyl sulfone and hydroxylated metabolites of polychlorinated biphenyls. In: Paasivirta J (ed) The Handbook of Environmental Chemistry, vol 3, part K. New Types of Persistent Halogenated Compounds Springer-Verlag, Berlin, pp 315–360.

Letcher RJ, Norstrom RJ, Muir DCG, Sandau CD, Koczanski K, Michaud R, De Guise S, Béland P (2000b) Methylsulfone polychlorinated biphenyl and 2,2-bis(chlorophenyl)-1,1-dichloroethylene metabolites in beluga whales (*Delphinapterus leucas*) from the St. Lawrence river estuary and western Hudson Bay, Canada. Environ Toxicol Chem 19:1378–1388.

Li H, Boon JP, Lewis WE, van den Berg M, Nyman M, Letcher RJ (2003) Hepatic microsomial cytochrome P450 enzyme activity in relation to *in vitro* metabolism/ inhibition of polychlorinated biphenyls and testosterone in Baltic grey seal (*Halichoerus grypus*). Environ Toxicol Chem 22:636–644.

Li YF, Bidleman TF, Barrie LA, McConnell LL (1998) Global hexachlorocyclohexane use trends and their impact on the arctic atmospheric environment. Geophys Res Lett 25:39–42.

Li YF, Macdonald RW, Jantunen LMM, Harner T, Bidleman TF, Strachan WMJ (2002) The transport of β-hexachlrohexane to the western Arctic ocean: a contrast to α-HCH. Sci Total Environ 291:229–246.

Lindström G, Wingfors H, Dam M, Bavel Bv (1999) Identification of 19 polybrominated diphenyl ethers (PBDEs) in long-finned pilot whale (*Globicephala melas*) from the Atlantic. Arch Environ Contam Toxicol 36:355–363.

Loganathan BG, Tanabe S, Tanaka H, Watanabe S, Miyazaki N, Amano M, Tatsukawa R (1990) Comparison of organochlorine residue levels in the striped dolphin from western north Pacific, 1978–79 and 1986. Mar Pollut Bull 21:435–439.

Lund B-O, Örberg J, Bergman Å, Larsson C, Bergman A, Bäcklin B-M, Håkansson H, Madej A, Brouwer A, Brunstrom B (1999) Chronic and reproductive toxicity of mixture of 15 methylsulfonyl-polychlorinated biphenyls and 3-methylsulfonyl-2,2-bis-(4-chlorophenyl)-1,1-dichloroethene in mink (*Mustela vison*). Environ Toxicol Chem 18: 292–298.

Lydersen C, Wolkers H, Severinsen T, Kleivane L, Nordøy ES, Skaare JU (2002) Blood is a poor substrate for monitoring pollution burdens in phocid seals. Sci Total Environ 292:193–203.

Macdonald RW, Barrie LA, Bidleman TF, Diamond ML, Gregor DJ, Semkin RG, Strachan WMJ, Li YF, Wania F, Alaee M, Alexeeva LB, Backus SM, Bailey R, Bewers JM, Gobeil C, Halsall CJ, Harner T, Hoff JT, Jantunen LMM, Lockhart WL, Mackay D, Muir DCG, Pudykiewicz J, Reimer KJ, Smith JN, Stern GA, Schroeder WH, Wagemann R, Yunker MB (2000) Contaminants in the Canadian Arctic: 5 years of progress in understanding sources, occurrence and pathways. Sci Total Environ 254: 93–234.

Mackay D, Wania F (1995) Transport of contaminants to the Arctic: partitioning, processes and models. Sci Total Environ 160–161:25–38.

Mandalakis M, Berresheim H, Stephanou EG (2003) Direct evidence for destruction of polychlorobiphenyls by OH radical in the subtropical troposphere. Environ Sci Technol 37:542–547.

Marsili L, Focardi S (1996) Organochlorine levels in subcutaneous blubber biopsies of fin whales (*Balaenoptera physalus*) and striped dolphins (*Stenella coeruleoalba*) from the Mediterranean Sea. Environ Pollut 91:1–9.

Marsili L, Focardi S (1997) Chlorinated hydrocarbon (HCB, DDTs and PCBs) levels in cetaceans stranded along the Italian coasts: an overview. Environ Monit Assess 45: 129–180.

Martineau D, Béland P, Desjardins C, Lagacé A (1987) Levels of organochlorine chemicals in tissues of beluga whales (*Delphinapterus leucas*) from the St. Lawrence Estuary, Québec, Canada. Arch Environ Contam Toxicol 16:137–147.

Martineau D, De Guise S, Fournier M, Shugart L, Girard C, Lagacé A, Béland P (1994) Pathology and toxicology of beluga whales from the St. Lawrence estuary, Québec, Canada. Past, present and future. Sci Total Environ 154:201–215.

Martineau D, Lemberger K, Dallaire A, Labelle P, Lipscomb TP, Michel P, Mikaelian I (2002) Cancer in wildlife, a case study: beluga from the St. Lawrence Estuary, Québec, Canada. Environ Health Perspect 110:1–8.

Massé R, Martineau D, Tremblay L, Béland P (1986) Concentrations and chromatographic profile of DDT metabolites and polychlorobiphenyl (PCB) residues in stranded beluga whales (*Delphinapterus leucas*) from the St. Lawrence estuary, Canada. Arch Environ Contam Toxicol 15:567–579.

Matkin CO, Ellis G, Olesiuk P, Saulitis E (1999) Association patterns and inferred genealogies of resident killer whales, *Orcinus orca*, in Prince William Sound, Alaska. Fish Bull 97: 900–919.

McDonald TA (2002) A perspective on the potential health risk of PBDEs. Chemosphere 46:745–755.

McKenzie C, Rogan E, Reid RJ, Wells DE (1997) Concentrations and patterns of organic contaminants in Atlantic white-sided dolphins (*Lagenorhyncus acutus*) from Irish and Scottish coastal waters. Environ Pollut 98:15–27.

McKinney MA, Arukwe A, Béland P, Martineau D, Dallaire A, Lair S, De Guise S, Lebeuf M, Letcher RJ (Submitted) Characterization and profiling of hepatic cytochromes P450 and phase II xenobiotic-metabolizing enzymes in beluga whales (*Delphinapterus* LEUCAS) from the St. Lawrence river estuary and the Canadian Arctic. Aquat. Toxicol 69:35–49.

Meerts IATM, Assink Y, Cenijn PH, van den Berg JHJ, Weijers BM, Bergman A, Koeman JH, Brouwer A (2002) Placental transfer of a hydroxylated polychlorinated biphenyl and effects on fetal and maternal thyroid hormone homeostasis in the rat. Toxicol Sci 68:361–371.

Meijer SN, Ockenden WA, Sweetman A, Breivik K, Grimalt JO, Jones KC (2003) Global distribution and budget of PCBs and HCB in background surface soils: implications for sources and environmental processes. Environ Sci Technol 37:667–672.

Minh TB, Watanabe M, Nakata H, Tanabe S, Jefferson TA (1999) Contamination by persistent organochlorines in small cetaceans from Hong Kong coastal waters. Mar Pollut Bull 39:383–392.

Minh TB, Watanabe M, Tanabe S, Miyazaki N, Jefferson TA, Prudente MS, Subramanian A, Karuppiah S (2000) Widespread contamination by *tris*(4-chlorophenyl)methane and *tris*(4-chlorophenyl)methanol in cetaceans from the North Pacific and Asian coastal waters. Environ Pollut 110:459–468.

Morris RJ, Law RJ, Allchin CR, Kelly CA, Fileman CF (1989) Metals and organochlorines in dolphins and porpoises of Cardigan Bay, West Wales. Mar Pollut Bull 20: 512–523.

Mos L, Ross PS (2002) Vitamin A physiology in the precocious harbour seal (*Phoca vitulina*): a tissue-based biomarker approach. Can J Zool 80: 1511–1519.

Muir DCG, Wagemann R, Grift NP, Norstrom RJ, Simon M, Lien J (1988) Organochlorine chemical and heavy metal contaminants in white-beaked dolphins (*Lagenorhyncus albirostris*) and pilot whales (*Globicephala melaena*) from the coast of Newfoundland, Canada. Arch Environ Contam Toxicol 17:613–629.

Muir DCG, Ford CA, Stewart REA, Smith TG, Addison RF, Zinck ME, Béland P (1990)

Organochlorine contaminants in beluga, *Delphinapterus leucas,* from Canadian waters. Can J Fish Aquat Sci 224:165–190.

Muir DCG, Ford CA, Grift NP, Stewart REA (1992) Organochlorine contaminants in narwhal *(Monodon monoceros)* from the Canadian arctic. Environ Pollut 75:307–316.

Muir DCG, Ford CA, Rosenberg B, Norstrom RJ, Simon M, Béland P (1996a) Persistent organochlorines in beluga whales *(Delphinapterus leucas)* from the St. Lawrence river estuary: I. Concentrations and patterns of specific PCBs, chlorinated pesticides and polychlorinated dibenzo-*p*-dioxins and dibenzofurans. Environ Pollut 93:219–234.

Muir DCG, Koczanski K, Rosenberg B, Béland P (1996b) Persistent organochlorines in beluga whales *(Delphinapterus leucas)* from the St. Lawrence river estuary: II. Temporal trends, 1982–1994. Environ Pollut 93:235–245.

Muir D, Braune B, DeMarch B, Norstrom R, Wagemann R, Lockhart L, Hargrave B, Bright D, Addison R, Payne J, Reimer K (1999a) Spatial and temporal trends and effects of contaminants in the Canadian Arctic marine ecosystem: a review. Sci Total Environ 230:83–144.

Muir DCG, Bidleman TF, Stern GA (1999b) New persistent and bioaccumulative chemicals in arctic air, water/snow, and biota. In: Kalhok S (ed) Synopsis of Research Conducted Under the 1997/1998 Northern Contaminant Program. Indian and Northern Affairs, Ottawa, Canada, pp 165–169.

Muir DCG, Riget F, Cleeman M, Skaare JU, Kleivane L, Nakata H, Dietz R, Severinsen T, Tanabe S (2000) Circumpolar trends of PCBs and organochlorine pesticides in the arctic marine environment inferred from levels in ringed seals. Environ Sci Technol 34:2431–2438.

Murk AJ, Leonards PEG, van Hattum B, Luit R, van der Weiden MEJ, Smit M (1998) Application of biomarkers for exposure and effect polyhalogenated aromatic hydrocarbons in naturally exposed European otters *(Lutra lutra)*. Environ Toxicol Pharmacol 6:91–102.

Norstrom RJ, Simon M (1990) Polychlorinated dibenzo-*p*-dioxins and dibenzofurans in marine mammals in the Canadian north. Environ Pollut 66:1–19.

Norstrom RJ, Muir DCG, Ford CA, Simon M, Macdonald CR, Béland P (1992) Indicators of P450 monooxygenase activities in beluga *(Delphinapterus leucas)* and narwhal *(Monodon monoceros)* from patterns of PCB, PCDD and PCDF accumulation. Mar Environ Res 34:267–272.

Norstrom RJ, Belikov SE, Born EW, Garner GW, Malone B, Olpinski S, Ramsay MA, Schliebe S, Stirling I, Stishov MS, Taylor MK, Wiig O (1998) Chlorinated hydrocarbon contaminants in polar bears from Eastern Russia, North America, Greenland, and Svalbard: biomonitoring of Arctic pollution. Arch Environ Contam Toxicol 35: 354–367

Nyman M, Bergknut M, Fant ML, Raunio H, Jestoi M, Bengs C, Murk A, Koistinen J, Bäckman C, Pelkonen O, Tysklind M, Hirvi T, Helle E (2003) Contaminant exposure and effects in Baltic ringed and grey seals as assessed by biomarkers. Mar Environ Res 55:73–99.

Ono M, Kannan K, Wakimoto T, Tatsukawa R (1987) Dibenzofurans a greater global pollutant than dioxins? Evidence from analyses of open ocean killer whale. Mar Pollut Bull 12:640–643.

Osterhaus ADME, Vedder J (1988) Identification of virus causing recent seal deaths. Nature (Lond) 335:20.

Parkinson A (1996) Biotransformation of xenobiotics. In: Klassen CD (ed) Casarett & Doull's Toxicology: The Basic Science of Poisons. McGraw-Hill, New York, pp 113–186.

Parsons ECM, Chan HM (2001) Organochlorine and trace element contamination in bottlenose dolphins (*Tursiops truncatus*) from the South China Sea. Mar Pollut Bull 42:780–786.

Prudente M, Tanabe S, Watanabe M, Subramanian A, Miyazaki N, Suarez P, Tatsukawa R (1997) Organochlorine contamination in some Odontoceti species from the north Pacific and Indian Ocean. Mar Environ Res 44:415–427.

Reddy ML, Reif JS, Bachand A, Ridgway SH (2001) Opportunities for using Navy marine mammals to explore associations between organochlorine contaminants and unfavorable effects on reproduction. Sci Total Environ 274:171–182.

Reeves RR, Smith BD, Crespo EA, di Sciara GN (2003) Dolphins, Whales and Porpoises: 2002–2010 Conservation Action Plan for the World's Cetaceans. IUCN/SSC Cetacean Specialist Group. IUCN, Gland, Swizerland and Cambridge, UK ix + 139 pp.

Reich S, Jimenez B, Marsili L, Hernández LM, Schurig V, González MJ (1999) Congener specific determination and enantiomeric ratios of chiral polychlorinated biphenyls in striped dolphins (*Stenella coeruleoalba*) from the Mediterranean Sea. Environ Sci Technol 33:1787–1793.

Reijnders PJH (1986) Reproductive failure in common seals feeding on fish from polluted coastal waters. Nature (Lond) 324:456–457.

Reijnders PJH (1994) Toxicokinetics of chlorobiphenyls and associated physiological responses in marine mammals with particular reference to their potential for ecotoxicological risk assessment. Sci Total Environ 154:229–236.

Ross P, De Swart R, Addison R, Van Loveren H, Vos J, Osterhaus A (1996) Contaminant-induced immunotoxicity in harbour seals: wildlife at risk? Toxicology 112:157–169.

Ross PS, Ellis GM, Ikonomou MG, Barrett-Lennard LG, Addison RF (2000) High PCB concentrations in free-ranging Pacific killer whales, *Orcinus orca:* effects of age, sex and dietary preference. Mar Pollut Bull 40:504–515.

Ruchel M (2001) Toxic dolphins. In: A Greenpeace investigation of persistent organic pollutants (POPs) in South Australian bottlenose dolphins. Greenpeace, Sydney, Australia.

Salata GG, Wade TL, Sericano JL, Davis JW, Brooks JM (1995) Analysis of Gulf of Mexico bottlenose dolphins for organochlorine pesticides and PCBs. Environ Pollut 88:167–175.

Schwacke LH, Voit EO, Hansen LJ, Wells RS, Mitchum GB, Hohn AA, Fair PA (2002) Probabilistic risk assessment of reproductive effects of polychlorinated biphenyls on bottlenose dolphins (*Tursiops truncatus*) from the southeast United States coast. Environ Toxicol Chem 21:2752–2764.

Simmonds MP, Johnston PA, French MC, Reeve R, Hutchinson JD (1994) Organochlorines and mercury in pilot whale blubber consumed by Faroe islanders. Sci Total Environ 149:97–111.

Smyth M, Berrow S, Nixon E, Rogan E (2000) Polychlorinated biphenyls and organochlorines in by-caught harbour porpoises *Phocoena phocoena* and common dolphins *Delphinus delphis* from Irish coastal waters. Biol Environ 100B:85–96.

Stern GA (1999) Temporal trends of organochlorine contaminants in southeastern Baffin beluga. In: Jensen J (ed) Synopsis of Research Conducted Under the 1997/1998 Northern Contaminants Program. Indian and Northern Affairs, Ottawa, Canada, pp 197–204.

Stern GA, Muir DCG, Segstro MD, Dietz R, Heide-Jørgensen MP (1994) PCB's and other organochlorine contaminants in white whales (*Delphinapterus leucas*) from West Greenland: variations with age and sex. Medd Grønl, Biosci 39:245–259.

Stern GA, Addison RF (1999) Temporal trends of organochlorines in southeast Baffin beluga and Holman ringed seals. In: Kalhok S (ed) Synopsis of research conducted under the 1998–1999 Northern contaminants program. Indian and Northern Affairs, Ottawa, Canada, pp 203–212.

Storelli MM, Marcotrigiano GO (2003) Levels and congener pattern of polychlorinated biphenyls in the blubber of Mediterranean bottlenose dolphins *Tursiops truncatus*. Environ Int 28:559–565.

Struntz WDJ, Kucklick JR, Schantz MM, Becker PR, McFee WE, Stolen MK (2004) Persistent organic pollutants in rough-toothed dolphins (*Steno bredanensis*) sampled during an unusual mass stranding event. Mar Pollut Bull 48:164–192.

Subramanian AN, Tanabe S, Tatsukawa R, Saito S, Miyazaki N (1987) Reduction in the testosterone levels by PCBs and DDE in Dall's porpoises of northwestern North Pacific. Mar Pollut Bull 18:643–646.

Subramanian AN, Tanabe S, Tatsukawa R (1988) Use of organochlorines as chemical tracers in determining some reproductive parameters in *Dalli*-type Dall's porpoise *Phocoenoides dalli*. Mar Environ Res 25:161–174.

Tanabe S (1988) PCB problems in the future: foresight from current knowledge. Environ Pollut 50:5–28.

Tanabe S, Mori T, Tatsukawa R (1983) Global pollution of marine mammals by PCBs, DDTs and HCHs (BHCs). Chemosphere 12:1269–1275.

Tanabe S, Loganathan BG, Subramanian AN, Tatsukawa R (1987a) Organochlorine residues in short-finned pilot whale. Possible use as tracers of biological parameters. Mar Pollut Bull 18:561–563.

Tanabe S, Kannan N, Subramanian AN, Watanabe S, Tatsukawa R (1987b) Highly toxic coplanar PCBs: occurrence, source, persistency and toxic implications to wildlife and humans. Environ Pollut 47:147–163.

Tanabe S, Watanabe S, Kan H, Tatsukawa R (1988) Capacity and mode of PCB metabolism in small cetaceans. Mar Mamm Sci 4:103–124.

Tanabe S, Subramanian AN, Ramesh A, Kumaran PL, Miyazaki N, Tatsukawa R (1993) Persistent organochlorine residues in dolphins from the Bay of Bengal, South India. Mar Pollut Bull 26:311–316.

Tanabe S, Kumaran PL, Iwata H, Tatsukawa R, Miyazaki N (1996) Enantiomeric ratios of α-hexachlorocyclohexane in blubber of small cetaceans. Mar Pollut Bull 32:27–32.

Tanabe S, Madhusree B, Öztürk AA, Tatsukawa R, Miyazaki N, Özdamar E, Aral O, Samsun O, Öztürk B (1997a) Isomer-specific analysis of polychlorinated biphenyls in harbour porpoise (*Phocoena phocoena*) from the Black Sea. Mar Pollut Bull 34:712–720.

Tanabe S, Madhusree B, Öztürk AA, Tatsukawa R, Miyazaki N, Özdamar E, Aral O, Samsun O, Özturk B (1997b) Persistent organochlorine residues in harbour porpoise (*Phocoena phocoena*) from the Black Sea. Mar Pollut Bull 34:338–347.

Teramitsu I, Yamamoto Y, Chiba I, Iwata H, Tanabe S, Fujise Y, Kazusaka A, Akahori

F, Fujita S (2000) Identification of novel cytochrome P450 1A genes from five marine mammal species. Aquat Toxicol 51:145–153.

Thomas PT, Hinsdill RD (1978) Effect of polychlorinated biphenyls on the immune responses of Rhesus monkeys and mice. Toxicol Appl Pharmacol 44:41–51.

Tilbury KL, Adams NG, Krone CA, Meador JP, Early G, Varanasi U (1999) Organochlorines in stranded pilot whales (*Globicephala melaena*) from the coast of Massachusetts. Arch Environ Contam Toxicol 37:125–134.

Tittlemier S, Borrell A, Duffe J, Duignan PJ, Fair P, Hall A, Hoekstra P, Kovacs KM, Krahn MM, Lebeuf M, Lydersen C, Muir D, O'Hara T, Olsson M, Pranschke J, Ross P, Siebert U, Stern G, Tanabe S, Norstrom R (2002) Global distribution of halogenated dimethyl bipyrroles in marine mammal blubber. Arch Environ Contam Toxicol 43:244–255.

Tomy G, Helm P (2003) New persistent chemicals of concern in an Eastern Arctic food web. In: Synopsis of research conducted under the 2001–2003 Northern Contaminants Program. Indian and Northern Affairs, Ottawa, Canada pp 367–370.

Troisi GM, Haraguchi K, Simmonds MP, Mason CF (1998) Methyl sulfone metabolites of polychlorinated biphenyls (PCBs) in cetaceans from Irish and Aegean Seas. Arch Environ Contam Toxicol 35:121–128.

Troisi GM, Haraguchi K, Kaydoo DS, Nyman M, Aguilar A, Borrell A, Siebert U, Mason CF (2001) Bioaccumulation of polychlorinated biphenyls (PCBs) and dichlorodiphenylethane (DDE) methyl sulfones in tissues of seal and dolphin morbillivirus epizootic victims. J Toxicol Environ Health Part A 62:1–8.

UNEP (United Nations Environment Program) (2001) Final act of the conference of plenipotentiaries on the Stockholm Convention on persistent organic pollutants. UNEP, Geneva, Switzerland.

van Bavel B, Sundelin E, Lillbäck J, Dam M, Lindström G (1999) Supercritical fluid extraction of polybrominated diphenyl ethers, PBDEs, from long-finned pilot whale (Globicephala melas) from the Atlantic. Organohalogen Compd 40:359–362.

van de Vijver KI, Hoff PT, Das K, van Dongen W, Esmans EL, Jauniaux T, Bouquegneau JM, Blust R, de Coen W (2003) Perfluorinated chemicals infiltrate ocean waters: link between exposure levels and stable isotope ratios in marine mammals. Environ Sci Technol 37:5545–5550.

van Hezik CM, Letcher RJ, de Geus H-J, Wester PG, Goksøyr A, Lewis WE, Boon JP (2001) Indications for the involvement of a CYP3A-like *iso*-enzyme in the metabolism of chlorobornane (Toxaphene®) congeners in seals from inhibition studies with liver microsomes. Aquat Toxicol 51:319–333.

van Scheppingen WB, Verhoeven AJIM, Mulder P, Addink MJ, Smeenk C (1996) Polychlorinated biphenyls, dibenzo-*p*-dioxins, and dibenzofurans in harbor porpoises (*Phocoena phocoena*) stranded on the Dutch coast between 1990 and 1993. Arch Environ Contam Toxicol 30:492–502.

Vetter W, Luckas B, Heidemann G, Skírnisson K (1996) Organochlorine residues in marine mammals from the northern hemisphere: a consideration of the composition of organochlorine residues in the blubber of marine mammals. Sci Total Environ 186: 29–39.

Vetter W, Klobes U, Luckas B (2001a) Distribution and levels of eight toxaphene congeners in different tissues of marine mammals, birds, and cod livers. Chemosphere 43: 611–621.

Vetter W, Scholz E, Gaus C, Müller JF, Haynes D (2001b) Anthropogenic and natural organohalogen compounds in blubber of dolphins and dugongs (*Dugong dugon*) from the Northeastern Australia. Arch Environ Contam Toxicol 41:221–231.

Vos JG, De Roij T (1972) Immunosuppressive activity of a polychlorinated biphenyl preparation on the humoral immune response in Guinea pigs. Toxicol Appl Pharmacol 21:549–555.

Wade TL, Chambers L, Gardinali PR, Sericano JL, Jackson TJ (1997) Toxaphene, PCB, DDT and chlordane analyses of beluga whale blubber. Chemosphere 34:1351–1357.

Watanabe M, Shimada T, Nakamura S, Nishiyama N, Yamashita N, Tanabe S, Tatsukawa R (1989) Specific profile of liver microsomal cytochrome P-450 in dolphin and whales. Mar Environ Res 27:51–65.

Weisbrod AV, Shea D, Moore MJ, Stegeman JJ (2000) Bioaccumulation patterns of polychlorinated biphenyls and chlorinated pesticides in Northwest Atlantic pilot whales. Environ Toxicol Chem 19:667–677.

Weisbrod AV, Shea D, Moore MJ, Stegeman JJ (2001) Species, tissue and gender-related organochlorine bioaccumulation in white-sided dolphins, pilot whales and their common prey in the Northwest Atlantic. Mar Environ Res 51:29–50.

Wells DE, Echarri I (1992) Determination of individual chlorobiphenyls (CBs), including non-ortho, and *mono*-ortho chloro substituted CBs in marine mammals from Scottish waters. Int J Environ Anal Chem 47:75–97.

Wells DE, Campbell LA, Ross HM, Thompson PM, Lockyer CH (1994) Organochlorine residues in harbour porpoise and bottlenose dolphins stranded on the coast of Scotland, 1988–1991. Sci Total Environ 151:77–99.

Wells RS (1999) Reproduction in wild bottlenose dolphins: overview of patterns observed during a long-term study. In: Duffield D, Tobeck T (eds) Bottlenose Dolphin Reproduction Workshop, Silver Springs, MD, USA. AZA Marine Mammal Taxon Advisory Group, San Diego, CA, pp 57–74.

Westgate AJ, Muir DCG, Gaskin DE, Kingsley MCS (1997) Concentrations and accumulation patterns of organochlorine contaminants in the blubber of harbour porpoises, *Phocoena phocoena,* from the coast of Newfoundland, the Gulf of St. Lawrence and the Bay of Fundy/Gulf of Maine. Environ Pollut 95:105–119.

White RD, Hahn ME, Lockhart WL, Stegeman JJ (1994) Catalytic and immunochemical characterization of hepatic microsomal cytochromes P450 in beluga whale (*Delphinapterus leucas*). Toxicol Appl Pharmacol 126:45–57.

White RD, Shea D, Schlezinger JJ, Hanh ME, Stegeman JJ (2000) In vitro metabolism of polychlorinated biphenyl congeners by beluga whale (*Delphinapterus leucas*) and pilot whale (*Globicephala melas*) and relationship to cytochrome P450 expression. Comp Biochem Physiol Part C 126:267–284.

Wiig Ø, Berg V, Gjertz I, Seagars DJ, Skaare JU (2000) Use of skin biopsies for assessing levels of organochlorines in walruses (*Odobenus rosmarus*). Polar Biol 23:272–278.

Wolkers H, Burkow IC, Lydersen C, Witkamp RF (2000) Chlorinated pesticide concentrations, with an emphasis on polychlorinated camphenes (toxaphenes), in relation to cytochrome P450 enzyme activities in harp seals (*Phoca groenlandica*) from the Barents Sea. Environ Toxicol Chem 19:1632–1637.

Ylitalo GM, Matkin CO, Buzitis J, Krahn MM, Jones LL, Rowles T, Stein JE (2001) Influence of life-history parameters on organochlorine concentrations in free-ranging killer whales (*Orcinus orca*) from Prince William Sound, AK. Sci Total Environ 281:183–203.

Yogui GT, Santos MCO, Montone RC (2003) Chlorinated pesticides and polychlorinated biphenyls in marine tucuxi dolphins (*Sotalia fluviatilis*) from the Cananéia estuary, southeastern Brazil. Sci Total Environ 312:67–78.

Zhou T, Ross DG, DeVito MJ, Crofton KM (2001) Effects of short-term *in vivo* exposure to polybrominated diphenyl ethers on thyroid hormones and hepatic enzyme activities in weanling rats. Toxicol Sci 61:76–82.

Manuscript received January 21; accepted January 24, 2004.

Environmental Contamination and Human Exposure to Lead in Brazil

Monica M.B. Paoliello and Eduardo M. De Capitani

Contents

I. Introduction 59
II. Production, Imports, and Exports 60
III. Bioaccumulation 61
    A. Terrestrial Plants 62
    B. Aquatic Animals 62
IV. Sources of Contamination and Standards for Human Exposure Control 66
    A. Environmental Exposure 66
    B. General Population Exposure 77
    C. Body Burden 81
    D. Occupational Exposure: Brazilian Legislation 88
Summary 91
References 92

I. Introduction

Despite the large use of lead in Brazil during the last two centuries, exposing directly or indirectly the general population living around industrial and mining areas, national scientific reports about toxic effects on humans and on the ecosystem are scarce and dispersed.

The oldest Brazilian study on the hazard of lead as an environmental contaminant for humans dates from 1880; published as a Ph.D. thesis presented at the University of Bahia, it described the risk of exposed workers (Castilho 1880, as reported by Spínola et al. 1980). In 1906, a study was published as a technical report to the municipality of Recife, Pernambuco, pointing out the risk of lead intoxication for the general population living in the city, through drinking water distributed through lead pipes (Azevedo 1906, as reported by Spínola et al. 1980). Since that time, mostly during the decades from 1930 to 1960, a dozen

---

Communicated by Lilia Albert.

M.M.B. Paoliello (✉)
Departamento de Patologia, Análises Clínicas e Toxicológicas, Centro de Ciências da Saúde, Universidade Estadual de Londrina; Avenida Robert Koch 60, 86038–440, Londrina, Paraná, Brasil.

E. M. De Capitani
Centro de Controle de Intoxicações, Hospital de Clínicas da Unicamp, Faculdade de Ciências Médicas, Universidade Estadual de Campinas; Cidade Universitária "Zeferino Vaz," caixa postal 6111, 13081–970, Campinas, São Paulo, Brasil.

descriptive studies on worker exposure to lead in industrial sectors such as primary smelting, battery production, and the printing industry were published mainly as academic theses or technical reports to governmental agencies (Spinola et al. 1980).

Except for the study by Azevedo (1906, as reported by Spínola et al. 1980), lead environmental contamination and its effects on the general population, especially on children, became a scientific and public health concern only in the 1980s, with the investigations done by a multidisciplinary group of researchers from Bahia. Despite these outstanding efforts, one cannot, for instance, compare data from recent studies on lead levels in children with older data due to the scarcity of publications.

Most of the studies discussed in this review were carried out in the State of São Paulo, the most industrially developed state in the country, and in Bahia, where the largest lead mine and refining facilities of Brazil were active until recently. Therefore, much of the information presented here must not be generalized or extrapolated to the rest of the country, where specific situations of lead contamination are still to be investigated.

This chapter is the first comprehensive review on environmental lead contamination ever published, presenting aspects of production and uses of lead in the country, and on legislation establishing limit values for some of the environmental compartments contaminated by lead, examining the few studies on reference values for lead in blood in general populations, and discussing data on human exposure around industrial and mining areas.

II. Production, Imports, and Exports

Brazil's lead ore reserves are around 900,000 t (Table 1), located in the states of Minas Gerais (67.3%), Paraná (18.6%), Rio Grande do Sul (7.9%), and Bahia (4.3%), and other states (1.9%), constituting 0.7% of the worldwide reserves (BRASIL 2001a). The only lead producing company in the country still operating is located in the Paracatu, state of Minas Gerais, after the closure of a larger company in Adrianópolis, state of Paraná, in 1995.

Table 1. Brazil's lead ore reserves in the year 2000.

State/location	Tons
Bahia/Boquira	37,594
Mato Grosso/Cáceres	15,984
Minas Gerais/Paracatú	596,506
Paraná/Adrianópolis e Cerro Azul	164,572
Rio Grande do Sul/Caçapava do Sul	70,035
São Paulo/Iporanga	1,152
Total	885,843

Source: BRASIL 2001a.

Table 2. Production of processed lead in Brazil from 1997 to 2000.

Tons	1997	1998	1999	2000
Total	8,729	7566	10,328	8,832

Source: BRASIL 2001a.

Brazilian production of processed lead decreased to 8,800 t in 2001, compared to 10,300 t in 1999 (Table 2).

The country imports semimanufactured lead mainly from Peru (62%), Venezuela (9%), China (7%), the United Kingdom (6%), and Argentina (5%). Imports increased about 25% in 2000 compared to 1999. Figure 1 shows the magnitude and trends of lead imports and lead-based semimanufactured goods from 1998 to 2001 (BRASIL 2002). Exports are mostly lead minerals and concentrated ore, being much smaller than imports (Table 3). Seventy-six percent (76%) of the lead consumed in Brazil is linked to the production of automotive lead acid batteries, lead oxides for pigments, welding processes, and ammunitions (BRASIL 2001a).

III. Bioaccumulation

High levels of metals in soil may be absorbed by plants and mobilized to surface and underground water. Contamination of plants by metals may occur directly from water or soil, and in general concentrations of each metal are low and kept into narrow limits to guarantee normal biological activity (Jordão 1999).

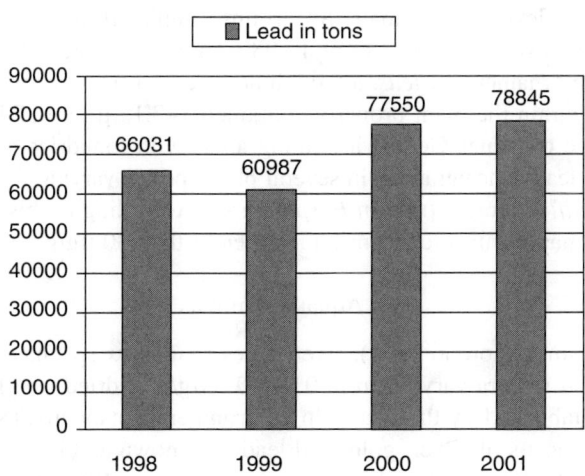

Fig. 1. Imports of lead and lead-based manufactured goods in Brazil, 1998 to 2001. *Source:* BRASIL 2002.

Table 3. Brazil's lead exports in tons from 1998 to 2000.

Description	1998	1999	2000
Pb minerals and concentrated ore	11,228	9,723	20,677
Semimanufactured and manufactured lead products	290	261	205
Chemical compounds	25	2	10
Total	11,543	9,986	20,892

Source: BRASIL 2002.

In aquatic organisms, lead absorption and accumulation from lead compounds present in water and sediments are influenced by several environmental factors such as temperature, salinity and pH, as well as the content of humic and alginic acids. In contaminated aquatic systems, a small fraction of lead is dissolved in water with the greater portion being strongly linked to the sediment (WHO 1989).

A. Terrestrial Plants

In a study carried out by the Environmental Agency of the State of São Paulo (CETESB) in 2002, in the surroundings of a lead acid battery recycling plant, in Bauru, State of São Paulo (see Section IV.A), samples of several kinds of vegetation were collected for lead content analysis. Table 4 presents the levels of lead in 15 samples collected in eight different locations. Lead determinations were carried out by graphite furnace atomic absorption spectrometry. For interpretation purposes, concentrations up to 7 ppm in dried weight were considered as maximum permitted values for human consumption, based on NOEL (no observed effect levels) in animals. Spearmint samples from one vegetable garden showed an average lead level of 12.98 μg/g and samples of common balm 6.54 μg/g. An ornamental *Ficus,* although not used for human consumption, that was located within the plant property contained 147.15 μg/g (CETESB 2002).

In the State of Minas Gerais, in mining areas with foundries present, Jordão (1999) found lead concentrations in several plant species varying from 0.8 μg/g in *Melinis minutiflora* to 1.4 μg/g in *Paspalum* sp. According to this author, toxicity to plants may occur at concentrations greater than 30 μg/g.

B. Aquatic Animals

In this same study (Jordão 1999), river fishes (*Astyanax* sp.) had levels of lead in muscles and viscera varying from 0.6 to 3.0 μg/g of dried weight, lower than the values established by the Brazilian environmental standards (8 μg/g).

Mineral deposits of silver, gold, and lead are known and have been explored since the 17th Century in the region of Vale do Ribeira (Ribeira de Iguape River in the extreme south of the State of São Paulo and east of the State of Paraná). In the beginning of the 20th Century, several mining companies were established in the region, contaminating the river and several of its affluents. A huge primary smelting and refining plant located at the margins of the river operated for more than 50 yrs, closing its doors in 1995. Several studies have been carried out by CETESB since the 1970s to monitor the concentrations of heavy metals in water, river sediments, and fishes. In 1996, assessing lead levels in the muscle of fishes regularly consumed by four small communities around the plant, CETESB found concentrations lower than those obtained in 1990 and 1991 (Table 5). Except for the dogfish species, all other species showed reduction in lead content. That decrease is probably related to the reduction in mining and refining activities between 1991 and 1995 (CETESB 1996).

The estuary systems of Santos and São Vicente, State of São Paulo, represent the most important examples of coastal environmental degradation by aquatic and atmospheric pollution of industrial origin in Brazil and in the world. The region shelters the largest port in Latin America (Port of Santos) and one of the biggest steel, chemical, and petrochemical industrial centers of the country, located in the municipality of Cubatão, which has been in operation since the 1950s. Wastes and sewage from the harbor and nearby urban centers add to the industrial pollutants and oil and toxic chemical spills from ships, resulting in a most desolate environmental picture (CETESB 2001b).

Not until 1984 did the process of degradation of the coastal ecosystems and the injurious effects of pollution start to be reversed, when an intensive program

Table 4. Concentrations of lead (dry weight basis) in vegetables cultivated around a lead acid battery plant in municipality of Bauru, State of São Paulo, Brazil.

Vegetable	Lead levels ($\mu g/g$) (ppm)
Mulberry	1.53
Spearmint	2.05
Kale	0.02
Grass	1.08
Parsley	0.66
Spearmint	2.11
Common balm	6.49
Lettuce	3.62
Spearmint	12.98
Ficus sp.	147.15
Mulberry	3.11
Spearmint	6.54
Manioc	2.08
Manioc	2.55
Manioc	2.91

Source: CETESB 2002.

Table 5. Lead content in muscle of several species of fishes from the Ribeira do Iguape river in 1996.

Fish species	Ribeira		Itaoca		Iporanga		Eldorado	
	n	Mean	n	Mean	n	Mean	n	Mean
Tapijara (*Hypostomus*), Aniã (*Hypostomus agna*)	10	0.27	7	0.22	13	0.28	—	—
Tapijara (*Hypostomus*), Cascudo (*Hypostomus*)	—	—	—	—	—	—	10	0.26
Mandi (*Pimelodus maculatus*)	9	0.15	13	0.24	12	0.23	4	0.60
Sagüirus (*Cyphocharax voga*)	—	—	7	0.33	3	0.13	9	0.27
Dogfish (*Oligosarcus hepsetus*)	3	0.57	4	0.29	9	0.33	2	0.25
Minnow (*Astyanax bimaculatus*)	10	0.34	—	—	3	0.42	—	—

Data are in µg/g.
Source: CETESB 1996.

for air, water, and soil pollution control in Cubatão was put into practice. Headed by CETESB, the Program for Environmental Quality Recovery of Cubatão was launched, emplacing systems for treatment of industrial effluents and resulting in a marked reduction in the pollution load. From the 1467 t/yr of heavy metals discharged by industries to the estuary in 1984, the program obtained a reduction of 97%, to only 44 tons in 1994 (CETESB 2001b).

In 1999, CETESB carried out a study assessing the lead content of some aquatic organisms living in the estuary system of Santos and São Vicente. Fishes, crustaceans, and mollusks were collected at 26 sampling points, totaling 161 specimens. Table 6 shows these results through the years 1979, 1989, and 1999. One can see that all results of 1999, when available, are much lower than those of 1979 and 1989, showing reduction in lead content of the species studied, despite the higher analytical detection limits of the measurements during 1979. As a matter of comparison, Brazilian regulatory agency establishes the limit value of 2.0 µg/g for lead in seafood. The highest average value was observed in oysters (0.41 ± 0.55 µg/g) collected in the estuary of Santos (CETESB 2001b).

In another study in the Bay of Santos and in the estuary of Santos and São Vicente, Boldrini and Pereira (1987) detected lead concentrations of 0.10 µg/g in muscle of fishes (carnivorous and omnivorous species) and 0.08–0.20 µg/g in viscera. A similar study was carried out between 1975 and 1983 in the Billings water reservoir on several species of fishes. This reservoir supplies most of the drinking water consumed in the southeast part of the State of São Paulo. Results showed maximum levels in viscera of 14.4 µg/g, including edible species (*Tilapia rendalis*), exceeding the established limit set by Brazilian law, which is 8 µg/g. Maximum lead level in muscle was 0.94 µg/g (Rocha et al. 1985).

Lead concentrations in *Sururu mytella falcata,* a commonly consumed mussel

Table 6. Lead levels (μg/g wet weight) in aquatic organisms in the estuary of Santos and São Vicente through the years of 1979, 1989, and 1999.

Organism	Rivers of Cubatão		Estuary of Santos			Estuary of São Vicente		Bay of Santos	
	1989	1999	1979	1989	1999	1979	1999	1979	1999
Fishes (muscle)	0.50 ± 0.16	<0.05	<0.20	0.42 ± 0.14	0.05	<0.20	<0.05	<0.20	—
Crabs	0.69 ± 0.04	0.32 ± 0.31	<0.20	0.59 ± 0.29	0.14 ± 0.11	<0.20	0.18 ± 0.25	<0.20	—
Mussels (muscle and viscera)	—	—	—	—	0.13 ± 0.14	<0.20	—	<0.20	0.13 ± 0.11
Oysters (muscle and viscera)	—	—	—	—	0.41 ± 0.55	<0.20	—	<0.20	—
Crab	—	—	—	—	0.08 ± 0.06	—	0.12 ± 0.11	—	—

Source: CETESB 2001b.

sampled in Mundau Lake, Maceió, State of Alagoas, varied between 0.72 and 9.0 μmol/g (Amaral 1989).

IV. Sources of Contamination and Standards for Human Exposure Control
A. Environmental Exposure

Air and Domestic Dust Thorton et al. (1995) estimated that one-third of daily exposure among urban populations comes from atmospheric sources. Mobile and stationary sources of lead tend to be concentrated in areas of high population density, or close to nonferrous metal foundries and refineries (ATSDR 1999).

A study done in Rio de Janeiro determined air lead levels near a lead acid battery recycling plant at distances varying from 25 to 500 m from the source of fumes and dust emissions. Results showed that lead levels varied from 0.07 to 183.3 $\mu g/m^3$. Considering the limit value of 1.5 $\mu g/m^3$ established by the U.S. Environmental Protection Agency (USEPA) for lead in the air, the authors observed that in 50% of the collected samples this limit was exceeded in the range of 25 m from the emission source (Quitério et al. 2001).

In that same study, Quitério et al. (2001) studied the lead levels in domestic dust at houses located near the battery recycling industry. From dust samples collected outside and inside the houses, 6% and 50%, respectively, exceeded the concentration of 1.500 $\mu g/m^2$ (limit level used in the study) at houses up to 50 m from the source. It should be noted that concentrations of lead in domestic dust vary considerably in different areas of the world and may be a significant source of exposure to the metal, specially for children. Several studies have associated lead concentrations in domestic dust and blood lead levels in children (Meyer et al. 1998; Trepka et al. 1997; Lanphear et al. 1998).

In February 2002, CETESB reported to the State Health Department a situation of lead contamination in the surroundings of a lead acid battery recycling plant in Bauru, State of São Paulo. The levels of lead in the atmosphere (total suspended particles) reached 37.3 $\mu g/m^3$ in 24 hr, with a geometric mean of 9.7 $\mu g/m^3$. Lead in house dust was also assessed, resulting in levels over 6000 ppm, whereas in a noncontaminated control area lead concentrations in domestic dust averaged 50 ppm (Freitas et al. 2002). Table 7 presents lead levels in air in areas near industrial emission sources in Brazil.

In an area near a closed primary lead smelting plant in the State of Paraná, De Capitani (personal communication, 2001) found concentrations of lead in domestic dust varying from 218 to 3268 ppm in houses where children presented blood lead levels above 20 μg/dL (24.1–37.8 μg/dL).

To monitor lead contamination in the spill areas of industries reprocessing the metal in Caçapava, State of São Paulo, lead in sedimentary dust was assessed in 1990–1993. Levels of lead in the sedimentary dust showed a high input of the metal into the ecosystem (Table 8). Levels of 200–1500 $kg/km^2/30$ d of sampling are usually found in sedimentary dust around lead foundries (Pompéia et al. 1993).

Table 7. Lead levels in total air in areas near industrial emission sources in Brazil.

Location	Source	Distance from source	Concentrations of Pb in air ($\mu g/m^3$)	Reference
Rio de Janeiro, State of Rio de Janeiro	Battery recycling industry	25–500 m	0.07–183	Quitério et al. 2001
Bauru, State of São Paulo	Battery recycling industry		Up to 37.3	Freitas et al. 2002
			9.7[a]	

[a]Average value.

The use of vitrified clays pottery to prepare and serve food is a very common practice in the State of Pernambuco, especially in Caruaru. In the vitrifying process, lead oxide (Pb_3O_4) is often used as the glaze. Lima et al. (2002) measured the content of lead in the vitrified pieces before and after burning. The content of lead released in the burning process varied from 211 to 13,660 mg/L, and the authors have estimated the loss of an average of 166 kg of lead per year to the environment.

Air Versus Leaded Gasoline International publications reveal a tendency for reduction of lead in the atmosphere in several countries, due mostly to the decrease in the use of lead (tetraethyl lead) as an additive to gasoline, besides the effort to control lead emissions from other sources. Such reductions have been reported in the United States, Canada, Germany, Norway, United Kingdom (WHO 1995) and Mexico (Flores and Albert 2004).

Until 1970, almost all gasoline used in the world contained lead, in many cases with concentrations above 0.4 g/L. Since the beginning of the 1970s, there has been a continuous movement toward controlling lead in fuels, guided in part by health concerns related to adverse effects on children and by the need for

Table 8. Levels of lead in sedimentary dust ($kg/km^2/30$ d) around a lead recycling plants in Caçapava, State of São Paulo, Brazil.

Sampled points	Sampling Date									
	1990	1992	Jan. 1993	Feb. 1993	Mar. 1993	Apr. 1993	May 1993	June 1993	July 1993	Aug. 1993	Sept. 1993
Industry 1	181.86	129.88	382.5	498.8	305.3	296.5	722.6	1033	1431	609.0	544.1
Industry 2	26.6	13.3	—	45.9	41.2	109.3	81.1	100.3	100.3	77.4	44.9

Source: Pompéia et al. 1993.

leadfree gasoline in cars using catalytic converters, in an international effort to reduce emissions of carbon monoxide, hydrocarbons, and nitrogen oxides (UNEP 1999).

In Brazil, the decrease in the use of lead in gasoline started as a consequence of the National Anhydrous Ethanol Program, a strategic program launched in 1975 in an attempt to diminish oil imports. Through this program, engine changes were made that aimed at using anhydrous ethanol as automotive fuel. As a consequence, a percentage of ethanol was also added to regular gasoline to boost its octane rating. Ethanol itself does not need lead to boost its octane rating, and therefore its lead content comes from the lead content of gasoline used in its denaturalization process (a process which uses 3% of gasoline).

Nevertheless, in 1979, a regulatory disposition (Resolution 14/79), issued by the National Council of Petroleum (CNP), defined gasoline type C (for ordinary, land, and water vehicles) as not containing tetraethyl lead, this being replaced by up to 22% of its volume by ethanol. The maximum addition of 0.8 ml/L tetraethyl lead would be restricted to gasoline types A and B (high octane rating), respectively, for exclusive use of the Air Force, and commercially distributed in the States of Acre, Rondônia, Amapá, and Roraima [Resolution 14/79 and Technical Regulation 04/79 of the National Council of Petroleum—Ministry of Mining and Energy (CNP-MME), September 4, 1979].

In 1982, a new resolution (Resolution 15/82) modified the former technical regulation of 1979, allowing a maximum of 0.8 ml/L tetraethyl lead also in ordinary gasoline, rescinding the decision of 1979 (Resolution 15/82 and Technical Regulation 04/79 Rev. I, CNP-MME, November 30, 1982). After those resolutions, there has not been another specific legislation that forbids the use of tetraethyl lead as a gasoline additive, but since 1993 its use became totally unnecessary with the requirement of using 22% ethanol as the additive in gasoline (7823/93 law). With this percentage of ethanol, tetraethyl lead negatively compromises the performance of engines. Meanwhile, data supplied by Romano et al. (1992), in a study carried out by CETESB, shows evidence that this percentage of ethanol, and the withdrawal of lead in gasoline, has probably been used since 1985. Although there is not a more updated specific legislation, data from this study done in the city of São Paulo (Tables 9–11) point to an evident reduction of lead in the atmosphere.

To follow up on the evolution of lead concentrations in the atmosphere of the metropolitan region of São Paulo, Romano et al. (1992) evaluated the levels of the metal in air during 1978 and 1987. They obtained air samples from several downtown locations, from residential, commercial, and industrial areas, and from a central park. The results of this study (see Tables 9–11) were obtained from total suspended particle samples. Air samples were collected during 24-hr periods every 6 d. Between 1978 and 1983, samples were analyzed by atomic absorption spectrometry, and samples from 1987 by X-ray fluorescence. Table 9 shows that air lead level in São Caetano, a highly industrialized region, in 1978 was over the standard limit value suggested by EPA (1.5 µg Pb/m^3) in two

Table 9. Quarterly averages of atmospheric lead in February–September 1978 (μg/m³).

Quarter	República (downtown) (n = 66)	São Caetano (industrial) (n = 72)	Pinheiros (residential) (n = 72)	Embu-Guaçú (rural) (n = 63)
Feb./Mar./Apr.	0.84	0.83	0.97	0.21
Mar./Apr./May	0.88	0.86	0.95	0.28
Apr./May/June	1.02	0.99	0.94	0.36
May/June/July	1.26	1.17	1.02	0.34
June/July/Aug.	1.42	1.60	1.20	0.28
July/Aug./Sept.	1.23	1.53	1.08	0.20

Source: Romano et al. 1992.

periods of sampling and that in Embu-Guaçú, a small rural neighbor, the levels stayed below that limit.

Table 12 shows that from 1978 to 1987 there was a reduction in air lead concentrations. In São Caetano and Pinheiros, a residential area, there were reductions of 72% and 82%, respectively. The changes observed in São Caetano (increase of 0.11 μg/m³) and Parque Dom Pedro (decrease of 0.24 μg/m³) are probably due to changes in the surrounding vehicle traffic. The authors also show that the use of ethanol in gasoline has steadily grown from 1979, and accordingly, the tetraethyl lead content in gasoline has decreased (from 0.145

Table 10. Quarterly averages of atmospheric lead in January—December 1983 (μg/m³).

Quarter	Ibirapuera (urban central park) (n = 53)	Parque Dom Pedro (urban central park) (n = 52)	Pinheiros (residential) (n = 57)	Osasco (industrial and residential) (n = 50)	São Caetano (industrial) (n = 55)
Jan./Feb./Mar.	0.12	0.26	0.25	0.12	0.31
Feb./Mar./Apr.	0.17	0.42	0.23	0.12	0.33
Mar./Apr./May	0.22	0.48	0.21	0.15	0.22
Apr./May/June	0.28	0.64	0.19	0.18	0.24
May/June/July	0.28	0.63	0.19	0.18	0.33
June/July/Aug.	0.27	0.50	0.17	0.20	0.37
July/Aug./Sept.	0.22	0.38	0.12	0.19	0.41
Aug./Sept./Oct.	0.17	0.22	0.09	0.20	0.33
Sept./Oct./Nov.	0.18	0.21	0.09	0.16	0.31
Oct./Nov./Dec.	0.16	0.19	0.09	0.13	0.25

Source: Romano et al. 1992.

Table 11. Quarterly averages of atmospheric lead in 1987 ($\mu g/m^3$).

Quarter	Ibirapuera ($n = 49$)	Parque Dom Pedro ($n = 46$)	Osasco ($n = 53$)	São Caetano ($n = 51$)
Jan./Feb./Mar.	0.14	0.22	0.18	0.37
Feb./Mar./Apr.	0.15	0.23	0.17	0.43
Mar./Apr./May	0.15	0.23	0.17	0.42
Apr./May/June	0.25	0.21	0.21	0.41
May/June/July	0.28	0.25	0.19	0.41
June/July/Aug.	0.33	0.34	0.20	0.46
July/Aug./Sept.	0.24	0.36	0.16	0.44
Aug./Sept./Oct.	0.20	0.32	0.14	0.37

Source: Romano et al. 1992.

ml/L in 1979 to 0.063 ml/L in 1988) (Table 13) (Romano et al. 1992). The reduction of 73.3% in the amount of tetraethyl lead (TEL) in gasoline between 1977 and 1988 (see Table 13) closely matches the reduction observed in air lead concentrations in the same period (see Table 12) (Romano et al. 1992).

Soil and Underground Water Natural concentrations of lead in soil are generally low, not exceeding 30 ppm in average (WHO 1995). Lead levels in urban soil can vary depending on the deposition rates and accumulation of atmospheric particles from anthropogenic sources.

In the State of São Paulo, CETESB has set reference values for metals in soils and underground waters (CETESB 2001a). Based on environmental studies of soils considered not contaminated, a reference value for lead was set as 17 mg/kg (Table 14).

Alert values represent a probable change in the natural background quality of the soil, alerting for an anthropogenic source of contamination. According to this definition, a value of 100 mg/kg of lead in soil is considered an alert value.

Table 12. Comparison of atmospheric lead levels in 1978, 1983, and 1987 ($\mu g/m^3$).

Sampling site	1978	1983	1987
São Caetano (industrial)	1.16	0.32	0.43
Pinheiros (residential)	1.03	0.18	—
Ibirapuera (urban central park)	—	0.24	0.23
Parque Dom Pedro (downtown central park)	—	0.51	0.27
Osasco (industrial/residential)	—	0.17	0.18

Source: Romano et al. 1992.

Table 13. Lead and ethanol content in automotive fuels in Brazil during 1979–1988.

Year	TEL (tetraethyl lead use in tons)	Average TEL content of gasoline (mL/L)	Average content of ethanol in gasoline (%)	Anhydrous ethanol consumption (megaliter)	Hydrated ethanol consumption (megaliter)
1977	8211	0.236	4.5	—	—
1978	—	—	8.5	—	—
1979	5160	0.145	15	—	—
1980	—	—	20	2.3	0.4
1981	3114	0.114	20	1.1	1.4
1982	—	—	20	2.0	1.7
1983	1809	0.087	20	2.2	3.0
1984	—	—	20	2.1	4.5
1985	2000	0.120	22	2.1	5.9
1986	—	—	22	2.4	8.2
1987	1044	0.060	22	2.1	8.8
1988	858	0.063	22	2.0	9.6

Source: Romano et al. 1992.

Intervention values are based on studies of risk assessment and have a corrective purpose. They indicate the existence of contamination of soils or underground waters, being used in the management of contaminated areas. Intervention values are set separately for agricultural, residential, and industrial soils (200, 350, and 1200 mg/kg, respectively) (see Table 14).

The range of variation of soil lead levels observed among several countries (see Table 14), may be explained by the use of different safety factors in the extrapolation of experimental data from animals to humans, in using or not using risk assessment criteria, and also by differences in local economic and environmental politics (CETESB 2001a).

In a study carried out in areas near an acid lead battery recycling plant in Bauru, State of São Paulo (air and domestic dust), a superficial soil layer of 20 cm depth was sampled from 44 points. At three points, lead levels were about 425 and 539 µg/g, and at four sampled points, the concentrations varied from 2.21 to 19.20 µg/g, all above the intervention values for residential and industrial soils (see Table 14) (CETESB 2002).

Table 15 presents soil lead levels in studies of areas close to industrial sources of lead emissions in Brazil. The importance of lead in soil as the preferential route of exposure for children must be emphasized. Several studies have already associated soil lead levels and child blood lead levels (Bjerre et al. 1993; Cook et al. 1993; Elhelu et al. 1995; Silvany-Neto et al. 1996; Lanphear et al. 1998; Murgueytio et al. 1998; Paoliello et al. 2002).

For underground waters, Table 16 presents the baseline values set for the State of São Paulo, comparing them with international values. The proposed

Table 14. Limit values for lead in soil (mg/kg) in the State of São Paulo, Brazil, and in other countries.

Place	Reference values	Alert values	Intervention values			SSL[a] (soil ingestion)	Multifunctionality	Trigger values (direct ingestion of soil)			
			Agriculture (APMax[b])	Residential	Industrial	Residential		Park	Residential	Park	Industrial
Brazil[c] (State of São Paulo)	17	100	200	350	1200	—	—	—	—	—	—
United States	—	—	—	—	—	400	—	—	—	—	—
Canada	—	—	375	500	1000	—	—	—	—	—	—
Holland	—	—	—	—	—	—	530	—	—	—	—
Germany	—	—	—	—	—	—	—	200	400	1000	2000

[a]Soil Screen Levels [Environmental Protection Agency (EPA)]; [b]Agricultural area [area of maximum protection (APMax)]; [c]Based on risk to the child.
Source: CETESB 2001a (modified).

Table 15. Levels of lead in soil (mg/kg) from areas near industrial emissions in Brazil.

Place, kind of industry	Soil lead levels	Reference
Adrianópolis, State of Paraná, inactive primary lead smelting plant	117–940	Cunha 2003
Bauru, State of São Paulo, acid lead battery recycling plant	<50–19,200	CETESB 2002
Santo Amaro da Purificação, State of Bahia, primary lead smelting plant		Silvany-Neto et al. 1996
in 1980	>10,000	
in 1985	>7,500	

level of intervention (10 µg/L or 0.01 mg/L) was based on a national regulation (Portaria 1.469 of 12/29/2000 from the Ministry of Health). The reference value generally accepted for underground waters is 1 µg/L (CETESB 2001a).

In the Bauru study, samples of undergound water were collected at 10 locations. In a well located just in front of the industry, lead levels were found to be 70 µg/L (see Table 16 for Brazilian intervention value). However, the authors think the water would have been contaminated by earth moving around the well performed by the industry at the time of sampling, and that the result should not be considered representative of underwater contamination by the recycling plant itself. Other wells around the plant showed lead levels between 0.011 and 0.025 mg/L (CETESB 2002).

Surface Water and Sediments In Brazil, the National Council for the Environment (CONAMA) established the maximum lead content allowed in drinking water with salinity lower than 0.50% at 0.03 mg/L. Saline water (salinity between 0.5% and 30%) and salty waters (salinity above 30%), which may be used for recreation, may have a maximum lead content of 0.01 mg/L (BRASIL 2001b) (Table 17).

Table 16. Reference values for lead in underground waters in the State of São Paulo compared to international values (µg/L).

Place	Intervention value	Trigger value	Multifunctionality
State of São Paulo (Brazil)	10	—	—
United States[a]	15	—	—
Canada (Quebec)[b]	100	—	—
Germany	—	25	—
Holland	—	—	75

[a]EPA (Environmental Protection Agency).
[b]Underground water severe contamination indicator.

Table 17. Maximum lead levels in surface waters, according to resolution 20 of CONAMA (1986), Brazil.

Drinking water		Saline water	Salty water
Class 2	Class 3	Class 5	Class 7
0.03 mg/L	0.05 mg/L	0.01 mg/L	0.01 mg/L

Source: BRASIL 2001b.

Cunha (2003), studying samples of surface water in affluents of the Ribeira de Iguape River from 1998 to 2000, found lead concentrations of less than 0.005 mg/L to a maximum of 0.006 mg/L, much lower than the maximum established by CONAMA (0.03 mg/L). Between 1978 and 1995, while lead mining and refining were still going on (see section III.B. Aquatic Animals), CETESB found lead concentrations varying from 0.004 to 0.23 mg/L in samples of surface waters of the Ribeira de Iguape River. Other authors, in 1986, had observed lead concentrations in surface waters of the same river varying from 0.01 to 2.75 mg/L (Eysink et al. 1998). After halting of the mining and refining activities, lead levels decreased drastically. Regarding sediments from the bottom of the Ribeira de Iguape River and its affluents, even after the conclusion of mining activities, lead levels still remain very high, varying from 30.8 and 968.8 µg/g. Reference value for sediments is 40 µg/g, according to Prates and Anderson (1977, as cited by Cunha 2003).

In Bauru, State of São Paulo, CETESB (2002) determined lead levels in surface waters at six sampling points. Table 18 shows that only one point had lead levels above reference values for surface water. For the assessment of sediment quality, that same study compared lead levels with the limit values established by the Canadian Environmental Agency in 1999 (Table 19). Brazil has not yet established maximum limits for lead in sediments. Table 18 shows that only two samples exceeded the TEL (threshold effect level), and another sample exceeded the PEL (probable effect level) for lead in sediments.

Table 18. Lead levels in surface waters and sediments (mg/L) in Bauru, State of São Paulo.

	Limit value (CONAMA, Brazil)	Reference value		Sampling points					
		TEL	PEL	1	2	3	4	5	6
Surface waters	0.03	—	—	<0.002	<0.002	0.005	0.004	0.002	0.57
Sediments	—	35.0	91.3	6.59	77.3, 76.7	1580	4.04	7.45	—

TEL, threshold effect level.
PEL, probable effect level (CEA 1999).
Source: CETESB 2002.

Table 19. Canadian criteria used to assess sediment quality regarding lead.[a]

Drinking water		Saline/salty water	
TEL[b]	PEL[c]	TEL[b]	PEL[c]
35 µg/g	91,3 µg/g	30,2 µg/g	112 µg/g

[a]Canadian Environment Agency (1999); [b]TEL (threshold effect level): threshold level below which adverse effects to biological organisms do not occur; [c]PEL (probable effect level): level of probable adverse effects to biological organisms (frequently associated with biological effects).

In the study developed by CETESB in 1999 in the estuary system of Santos and São Vicente, determinations of lead in surface waters and sediments found some concentrations above limits. Table 20 shows that 100% of the sediment samples were above detection limits of the method for lead, but only 11% presented concentrations above the TEL and 5% above the PEL. In relation to surface waters, only 9% presented levels above the limit established by Brazilian legislation (BRASIL 2001b). Table 21 shows chronological data regarding maximum lead concentrations found in sediments in that estuary system from 1979 to 1999, showing decreases in some areas and increases in others. The results of that study indicate the presence of lead contamination on sediments in all aquatic ecosystems of the region, especially near the main pollutant sources. The observed differences of results should be attributed to differences in sampling and analytical methods during that period (CETESB 2001b).

The historical analysis of lead concentration (dissolved phase) in waters of the rivers in Cubatão indicate that since 1983 limit values were exceeded only in 1990 (at Cubatão River and in a channel coming from a pollutant source)

Table 20. Lead in surface waters and sediments above the established limits in the estuary system of Santos and São Vicente in 1999.

Surface water			Sediments			
	Frequency (%)			Frequency (%)		
Number of samples	Above detected level	Above limit[a]	Number of samples	Above detected level	Above limits[b]	
					TEL[c]	PEL[d]
22	27	9	63	100	11	5

[a]CONAMA (1986); [b]Canadian Environment Agency (1999); [c]Threshold effect level; [d]Probable effect level.
Source: CETESB 2001b.

Table 21. Maximum lead levels in sediments (μg/g) in 1979, 1989, and 1999 in the estuary system of Santos and São Vicente, State of São Paulo.

	Santos (estuary)	São Vicente (estuary)	Bay of Santos	Rivers of Cubatão	Close to the pollutant source
1979	37.5	28.3	25.2	—	—
1989	26.6	—	—	41.8	—
1999	66.0	36.0	18.0	25.0	295

Source: CETESB 2001b.

and in 1992 (channel coming from a pollutant source). Lead concentrations found in surface water samples from the estuaries (Santos and São Vicente), and from the Bay of Santos, varied from 4 to 21 μg/L (average values) in 1979, and from less than 2 to 20 μg/L (absolute values) in 1999. Some of these results are above the limit value established by CONAMA for salty waters (CETESB 2001b).

In another study carried out at the Bay of Santos and at the estuaries of Santos and São Vicente, Boldrini and Pereira (1987) found lead concentrations in surface waters varying from 0.004 to 0.010 mg/L. In deep waters, the variation was from 0.006 to 0.021 mg/L. With relation to the average contents of lead in sediment, the variation was of 0.3 to 17.46 μg/g, with the estuary of Santos as the most contaminated. The average content obtained for all the region was 7.99 μg/g.

The State of Minas Gerais has huge deposits of iron, and some smelting, refining, and steel plants are present in the region. Jordão (1999), in a study carried out in Conselheiro Lafaiete and Ouro Branco, observed lead contamination in most of the surface water samples. Concentrations varied from 1.5 to 153 μg/L, which seem to be elevated when compared to concentrations of contaminated rivers in other localities. Lead concentrations in sediments were also relatively high, and the highest levels were found close to a mining area (60.3 μg/g). Baptista Neto et al. (2000) observed, in river sediments in Jurubatuba, State of Rio de Janeiro, concentrations varying from 5 to 123 ppm.

Yabe and Oliveira (1998), studying the river basin of Cambé, Municipality of Londrina, State of Paraná, assessed the concentration of several heavy metals in surface water. They found lead at 60.1 μg/L in the spring source of the stream, 69.01 μg/L at a point near a tannery, 288 μg/L upstream from a battery plant, 4584 μg/L in the spilling point of the same plant, 509.9 μg/L downstream from the plant, and 84.6 μg/L at Igapó Lake.

In Maceió, State of Alagoas, Amaral (1989) assessed lead levels in surface water from Lagoa do Mundau and found levels varying from 1.6 to 9.9 μmol/L and sediment values varying from 0.38 to 97 μmol/L. In a study between 1975 and 1983 of surface water of the Billings reservoir, which supplies most of the drinking water consumed in the southeast part of the State of São Paulo, levels did not exceed 0.02 mg/L (Rocha et al. 1985).

B. General Population Exposure

Foods and Beverages A regulatory act in 1998 (MS-685/1998) by the Ministry of Health of Brazil established the maximum permitted levels of contaminants in food, including the inorganic compounds and metal elements that constitute risks to human health. Those limit concentrations were established based on information from international regulatory statements, toxicological data, recommendations from the Codex Alimentarius, European Community, FDA (U.S. Food and Drug Administration), and other authorities. Table 22 presents the maximum permitted levels for lead in food, established by Brazilian legislation.

Lead in milk produced in an area around a secondary lead smelting plant in Caçapava, State of São Paulo (raw and pasteurized milk), was measured by Okada et al (1997). From 218 milk samples, 20% presented lead content above maximum limit (see Table 22), and the median value was 0.04 mg/L (0.01–0.20 mg/L) (Okada et al. 1997).

In Cananéia, State of São Paulo, lead content was measured in 69 samples of mangrove oysters (*Crassostrea brasiliana*), a great delicacy that has been grown in small sea farms in that region of the State for several decades. At the estuary of Cananéia, the contamination by heavy metal is potentially related to mining activity practiced in the upper Ribeira do Iguape River basin. Average lead levels found in the oysters was 0.08 mg/kg (maximum value, 0.17 mg/kg), below the maximum limit established by Brazilian legislation (Machado et al. 2002).

Table 22. Maximum permitted lead levels in food established in Brazilian legislation (MS-685/1998 from the Ministry of Health).

Food type	Maximum permitted levels for lead (mg/kg)
Oils, greases, refined emulsions	0.1
Caramels and candies	2.0
Cocoa bean (except cocoa butter and sweetened chocolate)	2.0
Sweetened chocolate	1.0
Dextrose (glucose)	2.0
Citric fruit juice	0.3
Fluid milk, ready to drink	0.05
Fish and fish products	2.0
Food for special purposes, specially prepared for sucklings and children under 3 years old	0.2
Cephalopod (edible parts)	2.0

A similar study regarding lead levels in bivalves was carried out in 1996–1997 at the estuary of Santos and São Vicente. The study was conducted from September 1996 to February 1997, in a period when high quantities of bivalves are consumed by the local community and by tourists as well. The results showed that the studied bivalves (*Crassostrea brasiliana, Perna perna,* and *Mytella falcata*) presented average values less than those established by the Brazilian legislation. However, the authors recommend continutation of the bivalve monitoring programs for evaluation of their quality and also of the environment (Pereira et al. 2002).

Sakuma et al. (1989) analyzed 294 samples of six different kinds of vegetables grown commercially in the Municipality of São Paulo, State of São Paulo, for lead, cadmium, and zinc content. Results showed low levels of all analyzed metals, demonstrating that vegetables did not represent a risk to the consumer, according to that preliminary study.

Lead and other metal levels in mineral waters produced in Brazil were studied by Maio et al. (2002). Sixty-nine analyzed samples showed the content of lead and other metals to be below the limits of quantification of the methods.

The content of lead in mixtures of mineral salt used as a cattle food supplement were determined in samples produced in the States of São Paulo, Paraná, Mato Grosso, and Mato Grosso do Sul. In São Paulo and Paraná, 30% of the samples were above the limit established by the North American National Research Council (1980), which is 30 ppm. In the States of Mato Grosso and Mato Grosso do Sul, 40% of the samples had lead content above that limit (Marçal et al. 2001b).

According to the Normative Instruction 42 (December 20, 1999) from the Ministry of Agriculture, the Program for Control of Residues in Meat (PCRC) established that, in bovines, the permitted maximum limit for lead is 2 ppm in muscle, liver, or kidney. According to Blood and Radostits (1991; cited by Marçal 2001a), blood lead levels in nonexposed bovines range from 5 to 25 µg/dL.

Aranha et al. (1994) determined lead levels in viscera of bovines from freezing storage plants kept under the Federal Inspection Service of several states of Brazil. The authors analyzed 317 samples, 61 of liver and 256 of kidneys. Average lead concentrations found were 0.12 and 0.13 ppm, respectively.

In 1994, in Caçapava, State of São Paulo, an environmental lead contamination incident led to the lead intoxication and death of 49 animals among cows, horses, pigs, and chickens. A secondary smelting plant was the source of the contamination. The river (Ribeirão dos Mudos) receiving industrial wastes from that plant showed values of lead in surface water up to eight times the limit established by Brazilian legal standards. Blood lead levels in 11 remaining bovines showed an average value of 60.6 ± 26.7 µg/dL (Marçal et al. 2001a). Kuno et al. (1999), investigating the same incident, analyzed lead in 66 samples of chicken blood, 53 samples from five properties near the lead smelting plant, and 13 from a control group located 10 km from the emission source. The determinations were carried out by graphite furnace atomic absorption spectrometry (AAS-GF). The means of blood lead concentrations in exposed animals varied from 14.1 to 34.3 µg/dL, whereas in the control group the mean was 4.3 µg/dL (Table 23).

In the same region of the State of São Paulo, Kuno et al. (1995) analyzed blood lead of 17 bovines from different properties around a lead scrap recycling plant. The mean blood lead was 102.4 µg/dL, indicating possible environmental contamination. As a control group, blood lead levels of animals from distant properties varied from 14.6 to 16.2 µg/dL. The same authors evaluated exposure in three bovines that were grazing 300 m from a lead recycling plant in Pindorama, State of São Paulo, Brazil, and found blood lead levels between 53.4 and 111.3 µg/dL, compared to 8.0 µg/dL from animals grazing 4 km from the source (Kuno et al. 1994).

Mazzeo (1984), investigating the death of 20 horses from a property situated near a lead recycling plant in the Paraíba River valley, State of São Paulo, found blood levels up to 67.0 µg/dL. According to the authors, limit values for lead in blood of horses might be between 33.0 and 47.0 µg/dL.

Drinking Water and Effluents The scientific literature shows that the lead body burden contribution from domestic drinking water is usually very small, almost negligible (Meyer et al. 1998). Table 24 compares the maximum permitted lead levels established by Brazilian legislation for drinking water with other institutions. According to CONAMA (BRASIL 2001b), the effluents of any industrial pollutant source can only be released into water bodies with a maximum lead value equal to 0.5 mg/L.

The only study available with results of lead concentrations in domestic tap water in Brazil was carried out by Cunha (2003) in Adrianópolis, State of Paraná, in 21 houses 200–1000 m distant from a primary lead smelting plant. Lead levels found varied from less than 0.005 to 0.008 mg/L.

Other Routes of Exposure Phytotherapeutic medicines have been a growing concern regarding metal contamination all over the world. With the objective of

Table 23. Blood lead level (mean ± SD) of chickens from Caçapava, State of São Paulo, according to distance from lead source.

Property	Distance from source (m)	Number of samples	Means ± SD (µg/dL)	Range (µg/dL)
1	400	16	27.3 ± 15.5	4.9–55.4
2	150	7	34.3 ± 23.1	14.6–77.7
3	400	10	14.1 ± 6.7	3.4–27.7
4	600	4	30.6 ± 24.7	13.0–66.1
5	300	16	25.3 ± 18.3	6.1–70.8
Control group	10,000	13	4.3 ± 2.3	<2.0–9.1

Source: Kuno et al. 1999.

Table 24. Maximum permitted values recommended for lead in drinking water by CONAMA, Brazil, and by other agencies.

Agency	Maximum permitted values (mg/L)
CONAMA (Brazil)	0.03
Germany (environmental standard)	0.04
USEPA	0.015
World Health Organization (WHO)	0.01
Statewide equal value:	
Arizona (U.S.)	0.02
Maine (U.S.)	0.02

evaluating lead and other metals in those kinds of medicines, Machado (2001) analyzed 130 samples of different preparations produced in Brasilia, DF, representing the 10 most heavily consumed products. During 1998 and 1999, the author collected and analyzed samples of *Ginkgo biloba* (15 samples), Espinheira Santa (*Maytemus ilicifolia*) (13 samples), Cascara buckthorn (*Ramnus spurshiana*) (14 samples), chlorella (5 samples), eggplant (13 samples), horse-chestnut (*Aesculus hippocastanus*) (15 samples), Brazilian ginseng (13 samples), *Centella asiática* (urban spadeleaf) (17 samples), guaraná (*Paullinia cupana*) (13 samples), and artichoke (12 samples). Lead was not detected in any of the artichoke, eggplant, or guaraná samples. High levels of lead, ranging from 153 to 1480 µg/g, were obtained in samples of horse-chestnut, whereas the maximum lead concentration found in the other preparations was 22 µg/g. It should be considered that the ingestion of toxic metals through the consumption of phytotherapeutic medicines may significantly contribute to the provisory tolerable weekly ingestion (PTWI) and may reach 442% of the PTWI when horse-chestnut is regularly ingested.

In the formulation of plastic materials used in food containers there are many additives, among them dyes and colored pigments, satisfying esthetic demand and protecting against deleterious effects from outside light. In 1982 and 1989, Garrido et al. (1991) analyzed 997 samples of dyes and pigments used in food containers to determine the content of lead and other metals. The now discontinued National Commission of Norms and Standards for Food had established, at that time, a maximum lead content in additives for plastic containers of 0.01%. Resolution 105 (May 19, 1999) from the Ministry of Health established the same percentage as a limit. From the analyzed samples, 5.4% showed lead levels above 0.01% (Table 25).

In another study, several kinds of school articles were analyzed by Garrido et al. (1990) for the content of lead and cadmium. A total of 205 samples of school articles were collected, with lead levels being very much higher than

Table 25. Lead levels in dyes and pigments for plastic food containers (mg/kg) in 1982–1989, Brazil.

Year	Number of samples	Maximum concentration found	Number and % of samples with Pb > 0.01% (100 mg/kg)
1982	82	758,377	2 (2.4)
1983	155	200	4 (2.6)
1984	123	n.d.	0 (0)
1985	150	15,243	13 (8.7)
1986	94	16,000	6 (6.4)
1987	50	12,500	4 (8.0)
1988	158	4,900	14 (8.9)
1989	185	13,200	11 (5.9)

n.d., not detected.
Source: Garrido et al. 1991 (modified).

those of cadmium (Table 26). Considering that that kind of article is not regulated by international agencies, a maximum permitted lead level of 90 mg/kg was used to compare the results of the study. That limit value was based on reference values for toys already established by the Brazilian Association for Technical Standardization (ABNT) and regulated by the INMETRO (a governmental institute in charge of controlling measurement instruments and products).

C. Body Burden

Lead absorption from environmental sources depends on the amount of the metal, its physical state, and chemical speciation. Absorption is also influenced by characteristics related to the subject such as age, physiological state, nutritional condition, and possibly genetic factors.

Table 26. Lead levels in school articles (mg/kg) obtained in Brazil.

Article	Number of samples	Minimum value	Maximum value
Eraser	37	0.06	249
Wax chalk	36	0.10	5,000
Ink	17	0.07	46
Fountain pen	39	0.08	13
Crayons	23	1.88	14
Modeling clay	38	0.50	12
Pencil	4	0.08	25
Colored glue	11	0.09	1.40

Source: Garrido et al. 1990 (modified).

Children Standards defining acceptable blood lead levels for children have changed remarkably in the past 30 years. In 1975, the Centers for Disease Control and Prevention (CDC), in Atlanta, GA (USA), defined maximum lead concentration in blood for children as 30 µg/dL. Previously, in the 1960s, the accepted level was 60 µg/dL; in the 1980s, that level decreased to 25 µg/dL, and in 1991, with evidence of the occurrence of late neurological adverse effects in children with levels as low as 10 µg/dL, the CDC and the Worldwide Health Organization adopted this value as action level. The high risk of lead absorption and of adverse health effects in children is due mostly to a natural tendency to hand–mouth behaviour when playing outdoors, a higher gastrointestinal rate of absorption related to normal physiological patterns, compared to adults, and an increased susceptibility to adverse effects, mainly in the central nervous system (CDC 1991).

In Brazil, there are no data available allowing a definition of a reference value for blood lead in children. Some Brazilian studies have evaluated the exposure of children living near lead smelting plants and mining areas. In Santo Amaro da Purificação, State of Bahia, between 1960 and 1993, a primary lead smelting plant affected the community of the areas close to the plant (Carvalho et al. 1984, 1985, 1995, 2003; Tavares et al. 1989; Silvany-Neto et al. 1989, 1996). Foundry scrap, containing about 2%–3% lead, was used to pave the home driveways and backyard patios. The local executive city authority used huge amounts of that scrap to pave many streets and public places of the city. In a study carried out in 1980, the average blood lead levels (geometric mean) of 555 children between 1 and 9 years old, living within 900 m of the plant, was 59.1 ± 25.0 µg/dL. In 1985, after the adoption of some controlling measures, the average level observed in a sample of 53 children was 36.9 ± 22.9 µg/dL. In 1998, using a sample of children from 1 to 4 years of age, the average blood lead level was 17.1 ± 7.3 µg/dL, with 88% of the results above 10 µg/dL and 32% above 20 µg/dL (Carvalho et al. 2003).

In the region of the upper Ribeira de Iguape River Valley (State of São Paulo and at east side of the State of Paraná), part of the work population and their families still continue to live in the contaminated area, despite closure of the smelting plant and mining activities in 1995. Paoliello (2002) assessed the level of lead exposure in 295 children aged 7–14 years and obtained median results of 11.25 and 4.40 µg/dL from children very close to the refinery and those in the mining areas, respectively. Those median values differed significantly from a nonexposed population living 40 km from the areas, which was 1.80 µg/dL; 59% of the children had blood lead levels ≥10 µg/dL and 12.8% were ≥20 µg/dL (Tables 27, 28).

Using logistic regression model analysis and defining 10 µg/dL of blood lead level as the cutoff value, some variables showed significant association with high blood lead levels: residential area close to the refinery [odds ratio (OR) = 10.38; 95% confidence interval (CI) = 4.86–23.25]; former father's occupational lead exposure (OR = 4.07; 95% CI = 1.82–9.24); and male gender (OR = 2.60; 95% CI = 1.24–5.62) (Table 29). Those results demonstrated that residual con-

Table 27. Comparison between blood lead levels in children living close to a lead refinery in former mining areas and control group at Vale do Ribeira, Brazil.

Variable	n	Median[a]	Values[a] Min.–Max.	Percentile[a] (25–75)	Value of P[b]
Dwelling area:					
Exposed population: mining area near refinery of PB[c]	94	11.25	1.80–37.80	6.60–14.00	<0.0001
Exposed population: mining area distant from the refinery of Pb[d]	201	4.40	1.80–29.40	3.00–6.40	
Control group[e]	39	1.80	1.80–8.20	1.80–1.80	

[a]µg/dL; [b]Value of P through test of Kruskal–Wallis; [c]Community of Vila Mota and Capelinha (rural area, County of Adrianópolis); [d]County of Adrianópolis (urban and rural area), County of Iporanga (rural area), and County of Ribeira (urban area); [e]County of Cerro Azul (urban area).
Source: Paoliello 2002.

tamination still remains even after the interruption of industrial activities. There are no data on blood lead levels in children when the refinery and the mining were still active (Paoliello et al. 2002).

In a cross-sectional design in Bauru, State of São Paulo, 624 children between 0 and 12 years living within 1000 m from a secondary smelting lead recycling plant, and 31 children living 11 km distant from the source, which constituted the control group, were assessed. Mean blood lead value in the exposed group was 9.28 µg/dL; 35.8% presented levels equal to or above 10 µg/

Table 28. Percentage of children by blood lead levels category in areas of former lead mining and primary smelting activities in upper Ribeira de Iguape river valley, Brazil.

Blood lead level (µg/dL)	Mining community near the lead refinery (n = 94)		Other mining communities (n = 201)		Total (n = 295)	
	n	%	n	%	n	%
<10	38	44.40	184	91.50	222	75.25
≥10	56	59.60	17	8.50	73	24.75
≥20	12	12.80	01	0.50	13	4.41

Source: Paoliello et al. 2002.

Table 29. Blood lead levels in children relative to some variables in a logistic regression model analysis.

Variables	Adjusted odds ratio	95% confidence interval
Sex		
Female	1.00	
Male	2.60	1.24–5.62
Age		
≤11 years	1.00	
>11 years	1.37	0.66–2.95
Residence area		
Far from the lead refinery[a]	1.00	
Close to the lead refinery[b]	10.38	4.86–23.25
Former dwelling at the refinery village		
No	1.00	
Yes	1.87	0.72–4.95
Daily consumption of milk		
No	1.00	
Yes	0.85	0.41–1.74
Consumption of vegetables from home backyard		
No	1.00	
Yes	1.38	0.60–3.26
Consumption of fruits from home backyard		
No	1.00	
Yes	1.51	0.70–3.26
Father with former occupational lead exposure		
No	1.00	
Yes	4.07	1.82–9.24

Source: Paoliello et al. 2002.

dL. In the logistic regression model analysis, defining 10 µg/dL as cutoff value, the following variables showed significant association with high blood lead levels: dwelling on nonpaved street (OR = 7.46; 95% CI = 4.60–12.10), distance from the emission source (less than 500 m from the plant) (OR = 2.42; 95% CI = 1.62–3.63), occupational lead exposure of a family member (OR = 1.52; 95% CI = 1.13–2.03), and habit of playing on the ground (OR = 1.55; 95% CI = 1.03–2.31).

Santos Filho et al. (1993) found mean blood lead levels of 17.8 ± 5.8 µg/dL in 251 children between 1 and 10 years old living at the margins of the Cubatão River. Also in Cubatão, Azevedo et al. (1989) studied 199 children between 4 and 5 years, from 10 public schools, in 1983. Using flame AAS (Atomic Absorption Spectrometry) average values obtained varied from 5.02 µg/dL to 18.51 µg/dL (Table 30).

Table 30. Blood lead levels in children living in Cubatão, State of São Paulo, Brazil, in 1983.

School	n	Mean	Standard deviation	Amplitude of variation
1	48	11.14	5.35	3.8–25.6
2	37	7.32	4.09	0.6–15.1
3	15	18.51	8.82	0.6–35.7
4	22	5.02	3.68	0.6–35.7
5	20	10.18	4.21	5.5–19.5
6	10	10.25	5.86	2.8–22.2
7	15	6.53	5.00	1.0–16.6
8	19	5.83	2.46	0.7–12.8
9	6	8.43	2.51	5.5–11.7
10	7	8.43	2.51	5.5–11.7

Source: Azevedo et al. 1989.

Table 31 compares blood lead level data obtained in children living around industrial sources of lead in Brazil with data obtained in other countries, and Table 32 compares results from mining areas in Brazil and in other countries.

Adults It was demonstrated, through several studies in different populations, that factors such as age, sex, ethnic group, food habits, consumption of alcohol, smoking, hobbies, season of the year, dwelling area and geographic location cause interference in adult blood lead levels. Therefore, it is not possible to present international reference values for lead in adult blood. Besides that, sources of lead exposure are always changing in location and intensity with time, making reference values a temporary definition (Gerharsson et al. 1996).

Table 33 presents reference values obtained by Paoliello et al. (2001) in a nonexposed population in Londrina, State of Paraná. Table 34 compares the two studies on reference values carried out in Brazil with other countries. The study of Paoliello et al. (2001) was carried out in a poor industrialized city of approximately 490,000 inhabitants, with the economy based mainly on services, especially in areas such as health and trade. The study of Fernícola and Azevedo (1981) was carried out in São Paulo, the largest industrialized urban center in the country.

Paoliello et al. (2003) evaluated 350 subjects living in the proximity of the same lead refinery described above in the Municipality of Adrianópolis, State of Paraná. Logistic regression models were used to correlate some independent variables to blood lead levels and to assess the specific effect of each adjusted variable by the others. The blood lead level cutoff used was 14 µg/dL; this value was chosen because it was the superior limit of the reference interval obtained

Table 31. Means of blood lead levels (μg/dL) in children living near industrial sources in Brazil and other countries.

Location	Source	Proximity to source	Age (yr)	Exposed n	Exposed PbS	Control n	Control PbS	Reference
Santo Amaro da Purificação, Bahia, Brazil	Primary lead smelting plant	To 900 m	1–9	555	59.1[a] ± 25.0			Carvalho et al. 1985
Santo Amaro da Purificação, Bahia, Brazil	Primary lead smelting plant	To 900 m	1–9	53	36.9[a] ± 22.9			Silvany-Neto et al. 1989
Santo Amaro da Purificação, Bahia, Brazil	Primary lead smelting plant	<1 km	1–4	47	17.1[a] ± 7.3			Carvalho et al. 2003
Adrianópolis, Paraná, Brazil	Primary lead smelting plant	500 m–1.5 km	7–14	94	11.25[b]	39	1.8[b]	Paoliello et al. 2002
Bauru, State of São Paulo, Brazil	Secondary lead smelting and battery recycling plant industry	To 1 km	0–14	825	9.28[a]	31	<5.0	Freitas et al. 2002
At the margins of Cubatão River, São Paulo, Brazil	Steel plants, petrochemical		1–10	251	17.8[a] ± 5.8			Santos Filho et al. 1993
North of France	Metal foundry		8–12	200	3.97[c]	200	3.06[c]	Leroyer et al. 2000
Montevideo, Uruguay	Metal foundry	2 km	To 14	49	11.8	34	10	Mañay and Cousillas 1997
Antofagasta, Chile	Lead storage facilities	400 m	To 2	486	8.7[c]	75	4.22	Supúlveda et al. 2000
Mumbay, India	Industrial area		6–10	21	14.4[c]			Raghunath et al. 1999
Poland	Industrial area		7	431	7.94[c]			Zejda et al. 1995

[a]Mean; [b]Median; [c]Geometric mean.

Table 32. Means of blood lead levels (μg/dL) of children living around mining areas.

Location	n	Age	Level of Pbs (exposed)	Level of PbS (control)	Reference
Alto Vale do Ribeira, Brazil	210	7–14 yr	4.4[b]	1.8[b]	Paoliello et al. 2002
Falun, Sweden	49	0.7–7.4 yr	3.1		Bjerre et al. 1993
Leadville, Colorado	239	6–7 mon	10.1		Cook et al. 1993
Hettstedt, Germany	527	5–14 yr[a]	3.8[a]		Trepka et al. 1997
Hettstedt, Germany	418	5–14 yr[a]	3.5[a]		Meyer et al. 1999
Big River Mine Tailings, Missouri, USA	226	6–9 mon	6.52	3.43	Murgueytio et al. 1998
Sala, Sweden	202	1–5 yr	2.2	2.1	Berglund et al. 2000

[a]Geometric mean.
[b]Median.
Source: Paoliello and Decapitani 2003.

in previous studies (Paoliello et al. 2001). Five of the variables studied were independently associated with high blood lead levels in adults: residential area, gender, former dwelling at the refinery village, smoking habits, and consumption of fruits from home backyard (Table 35).

In the study of Kuno et al. (1994) in Pindorama, in an area around a lead

Table 33. Blood lead levels (μg/dL) in healthy adult individuals, Londrina, State of Paraná, by gender.

	Sex		
Statistic	Female ($n = 301$)	Male ($n = 219$)	Total ($n = 520$)
Minimum value	1.2	1.2	1.2
1st quartile	3.6	3.5	3.6
Median	5.7	5.6	5.7
3rd quartile	8.1	9.0	8.3
Maximum value	23.0	23.0	23.0
Geometric mean	5.4	5.6	5.5
Confidence interval (95%)	6.46 ± 0.40	6.62 ± 0.49	6.52 ± 0.30
Experimental range	1.20–23.0	1.20–23.0	1.20–23.0
Reference value	1.20–13.49	1.20–14.04	1.20–13.72
Uncertainty range	13.49–23.0	14.04–23.0	13.72–23.0

Source: Paolielo et al. 2001.

Table 34. Reference values for lead in blood (μg/dL) in adult population of locations compared to São Paulo and Londrina, Brazil.

Place	n	Average plumbemia	Standard deviation	Reference
São Paulo, São Paulo, Brazil	63 37	14.2 (men) 9.3 (women)	4.3	Fernícola and Azevedo 1981
Londrina, Paraná, Brazil	520	5.5[a] (total) 5.6[a] (men) 5.4[a] (women)		Paoliello et al. 2001
Italy (several places)	2,861 3,806	15.30[b] (men) 10.0[b] (women)		Morisi et al. 1989
Italy	203	13.8	4	Apostoli and Alessio 1990
Lombardia region, north of Italy	959	15.77		Minoia et al. 1990
Luca, Italy	299	8.4	4.1	Montesanti et al. 1995
Denmark	100	5.6 (men) 4.6 (women)	2.7 2.9	Grandjean 1992 (cited in Gerhardsson et al. 1996)
Taiwan, China	5,913 5,913	7.0[b] 8.28	5.39	Liou et al. 1996
Korea (several regions)	316 209	6.36[a] (men) 5.09[a] (women)	1.44 1.45	Yang et al. 1996
China	202	5.67[a] (women)		Zhang et al. 1997
Japan	72	3.21[a] (women)		Zhang et al. 1997
District of Florence, Italy	2,330 (total)	9.35[b] (men) 6.25[b] (women)		Donni et al. 1998

[a]Geometric mean.
[b]Median.
Source: Paoliello and Chasin 2001.

recycling plant, blood lead levels of 15 adults varied according to the distance from the plant, from mean values of 4.84 to 22.2 μg/dL (Table 36).

The only study of blood lead levels in pregnancy was done by Moura (1996) with 38 pregnant women from Rio de Janeiro. To analyze the data, lead results were adjusted by the values of red cells count in the blood, to control for the influence of physiological anemia during pregnancy. A statistically significant increase in blood lead levels was observed from the first to second and third trimester of the pregnancy, 5.1 μg/dL, 5.9 μg/dL, and 8.25 μg/dL, respectively.

Table 35. Blood lead levels in adults relative to variables studied in a logistic regression model analysis.

Variable	Adjusted odds ratio	95% confidence interval
Sex		
Female	1.00	
Male	18.25	5.41–62.35
Residential area		
Far from lead refinery	1.00	
Near lead refinery	7.27	2.61–20.24
Former dwelling at the refinery village		
No	1.00	
Yes	5.43	1.89–15.60
Smoking habit		
No	1.00	
Yes	4.24	1.44–12.49
Consumption of fruits from home backyard		
No	1.00	
Yes	3.63	1.32– 9.98
Alcohol consumption		
No	1.00	
Yes	0.45	0.13– 1.48
Daily consumption of milk		
No	1.00	
Yes	1.13	0.22– 5.92
Age		
15–34 yr	1.00	
Over 34 yr	0.69	0.23– 2.08

Source: Paoliello et al. 2003.

Table 36. Means of lead in blood in adults living near a lead recycling plant in Pindorama, State of São Paulo, relative to distance from source.

Number of samples	Distance from the source (m)	Average of lead in blood (μg/dL)
5	600–800	21.22
8	1500–1600	8.15
2	2000	4.85

Source: Kuno et al. 1994.

D. Occupational Exposure: Brazilian Legislation

The potential risks to occupational exposure in foundries and primary or secondary lead refining are well known. There are numerous Brazilian publications regarding the occupational exposure to lead, but, those data are not included in the scope of this review. Nevertheless, some limits established by Brazilian legislation for the occupational exposure are presented in Table 37, which presents the limits for lead in the workplace atmosphere established by the Brazilian legislation and by different authorities. Although most of the occupational exposure standards are based only on lead concentrations in the air, that route of exposure does not reflect the total daily exposure of the workers, which includes lead exposure through food, water, alcoholic beverages, and dust (WHO 1995). Table 38 presents the limit values for lead in blood established for occupational exposure in Brazil and in other countries.

In Brazil, according to a Regulatory Act (NR7 from the Ministry of Labor), the laboratory parameters for the biological monitoring of inorganic lead exposure are lead in blood, δ-aminolevulinic acid in urine (ALAU) or zincoprotoporphyrin (ZPP) in blood. Reference values and maximum permitted biological values are, respectively, up to 40 µg/dL and 60 µg/dL for lead in blood, to 4.5 mg/g of creatinine and 10 mg/g of creatinine for ALAU, and to 40 µg/dL and 100 µg/dL for ZPP in blood. For tetraethyl lead, the biological indicator of

Table 37. Limits of exposure to lead in the work environment atmosphere, according to Brazilian legislation and other agencies.

Authority	Recommended level
BRAZIL, LTs (mg/m^3)	0.1
ACGIH TLVs TWA (mg/m^3)	0.05
OSHA PELs TWA (mg/m^3)	0.05
NIOSH RELs TWA (mg/m^3)	<0.1[a]
DFG TWA (mg/m^3)	0.1
Carcinogenic category	EPA-B2
	IARC-2B
	TLV-A3

ACGIH, American Conference of Governmental Industrial Hygienists; OSHA, Occupational Safety and Health Administration; NIOSH, National Institute for Occupational Safety and Health; IARC, International Agency for Research on Cancer; DFG, Deutsche Forschungsgemeinschaft (German Research Foundation); TLVs, threshold limit values; PELs, permissible exposure limits; RELs, recommended exposure limits, LTs, tolerable limits; TWA, time-weighted exposure concentration; EPA-B, probable human carcinogenic; IARC-2B and TLV-A3, animal carcinogenic; epidemiological data do not reveal human carcinogenesis.
[a]Levels of lead in blood <0.06 mg/100 g.
Source: Paoliello and De Capitani 2003.

Table 38. Limits of lead in blood for occupational exposure established in Brazil and in other countries.

Country	Maximum plumbemia (µg/dL)
Men:	
South Africa	80
Canada, European Community, France, Germany, Greece, Ireland, Italy, Luxemburg, Spain, Thailand, United Kingdom	70
Brazil, Israel, Japan, Morocco, Holland, Peru	60
Australia, Belgium, Denmark, Finland, Norway, Sweden, United States of America	50
Women:	
South Africa, United Kingdom	40
Germany, Israel, Norway, Sweden	30
Australia	20

Source: Paoliello and De Capitani 2003.

exposure is lead in urine; the reference value is to 50 µg/g of creatinine, and the maximum permitted biological index is equal to 100 µg/dL.

Summary

Adverse effects caused by environmental lead pollution are well recognized. Being a widespread agent in the environment and a major harmful element to organic systems, mostly to children, lead has been investigated all over the world, aiming to improve measures regarding its control. The purpose of this chapter is to present a review of the situation of production, uses, assessment of exposure, and adverse effects from environmental lead contamination in Brazil. It also presents aspects of Brazilian legislation setting up maximum permissible levels of lead in several environmental compartments such as surface and drinking water, soils, sediment, urban air, and also in commercially sold food, vegetables, fish, and meat, in an effort to control industrial emissions. Epidemiological investigations on children's lead exposure around industrial and mining areas are revised, showing that many situations where lead contamination is potentially present still need to be addressed by governmental agencies. In Brazil, lead was withdrawn from gasoline by the end of the 1980s, and the last lead mining and primary smelting plant was closed in 1995, leaving residual environmental lead contamination, which has recently been investigated using a multidisciplinary approach. Nevertheless, there are hundreds of small secondary battery recycling plants all over the country, running smelting facilities that produce local urban areas of lead contamination.

References

Amaral OLC (1989) Níveis de concentração de cádmio, chumbo e mercúrio em águas, sedimentos e organismos da Lagoa de Mundau, Maceió. Mestrado, Pontifícia Universidade Católica do Rio de Janeiro, Rio de Janeiro.

Apostoli P, Alessio L (1990) Valori di referimento e controllo dei fattori di variabilitá. In: La Promazione della qualitá dei dati nel monitoraggio biológico. Moderna, Itália, pp. 111–127.

Aranha S, Nishikawa AM, Taka T, Salioni EMC (1994) Níveis de cádmio e chumbo em fígado e rins de bovinos. Rev Inst Adolfo Lutz 54(1):16–20.

ATSDR (1999) Toxicological profile for lead. ATSDR–Agency for Toxic Substances and Disease Registry, Atlanta, GA.

Azevedo FA, Queiroz IR, Kuno R, Bastian EYO, Maluf CB, Campos AEM, Diniz K (1989) Avaliação tóxico-epidemiológica da exposição ambiental da população infantil do Município de Cubatão (Sp-Brasil) a metais pesados: chumbo e mercúrio. Rev Soc Bras Toxicol 2(1):25–32.

Baptista Neto JA, Smith BJ, Mcallister JJ (2000) Heavy metal concentrations in surface sediments in a nearshore environment, Jurujuba Sound, Southeast Brazil. Environ Pollut 109:1–9.

Bjerre B, Berglund M, Harsbo K, Hellman B (1993) Blood lead concentrations of Swedish preschool children in a community with high lead levels from mine waste in soil and dust. Scand J Environ Health 19:154–161.

Boldrini CV, Pereira DN (1987) Metais pesados na Baía de Santos e Estuários de Santos e São Vicente—bioacumulação. Ambiente 1(3):118–127.

BRASIL (2001a) Ministério de Minas e Energia. Departamento Nacional de Produção Mineral. www.dnpm.gov.br/dnpm_legis/amb2001.html. Acesso em 5 de Janeiro, 2003.

BRASIL (2001b) Conselho Nacional do Meio Ambiente. Resolução CONAMA n° 20: Estabelece a classificação das águas e os níveis de qualidade exigidos. http://www.mma .gov.br/port/conama/frlegis.html. Acesso em 02 de Maio, 2001.

BRASIL (2002) Ministério de Minas e Energia.Perfil Nacional da Gestão de Substâncias Químicas. http://www.mma.gov.br/port/sqa/copasq/doc/pt/conasq.pdf. Acesso em 5 de Janeiro, 2003.

Carvalho FM, Barreto ML, Silvany-Netto AM, Waldron HA, Tavares TM (1984) Multiple causes of anaemia among children living near a lead smelter in Brazil. Sci Total Environ 35:71–84.

Carvalho FM, Silvany-Netto AM, Tavares TM, Lima MEC, Waldron HA (1985) Lead poisoning among children from Santo Amaro, Brazil. Pan Am Health Organ Bull 19(2):165–175.

Carvalho FM, Silvany-Netto AM, Barbosa AC, Cotrim ARL, Tavares TM (1995) Erythrocyte protoporphyrin versus blood lead: relationship with iron status among children exposed to gross environmental pollution. Environ Res 71:11–15.

Carvalho FM, Silvany Neto AM, Tavares TM, Costa ACA, Chaves CR, Nascimento LD, Reis MA (2003) Blood lead levels in children and environmental legacy of a lead foundry in Brazil. Pan Am J Public Health 13(1):19–23.

CDC (1991) Preventing lead poisoning in children. U.S. Centers for Disease Control and Prevention, U.S. Dept of Health and Human Services, Atlanta, GA.

CETESB (1996) Avaliação da qualidade do Rio Ribeira de Iguape. Relatório Complementar. CETESB–Companhia de Tecnologia de Saneamento Ambiental, São Paulo (BR).

CETESB (2001a) Relatório de estabelecimento de valores orientadores para solos e águas subterrâneas no Estado de São Paulo. www.cetesb.sp.gov.br/Solo/solo_geral.asp. Acesso em 16 de Novembro, 2003.

CETESB (2001b). Sistema Estuarino de Santos e São Vicente (CD-ROM). CETESB–Companhia de Tecnologia de Saneamento Ambiental, PROCOP-Programa de Controle de Poluição. Relatório Técnico, ag2001. Governo do Estado de São Paulo. Secretaria do Meio Ambiente, São Paulo, Brazil.

CETESB (2002) Avaliação confirmatória da contaminação ambiental da área de influência da indústria "Acumuladores AJAX Ltda-Recuperadora de chumbo" em Bauru-SP. CETESB–Companhia de Tecnologia de Saneamento Ambienta. Relatório v 1 Informação técnica n° 018/ECC/EQQ/EQS/ERQ/02. Diretoria de Recursos Híbridos e Engenharia Ambiental. Diretoria de Controle de Poluição Ambiental. Diretoria de Desenvolvimento e Transferência de Tecnologia. São Paulo, Brazil.

Cook M, Chappell WR, Hoffman RE, Mangione EJ (1993) Assessment of blood lead levels in children living in a historic mining and smelting community. Am J Epidemiol 137:447–455.

Cunha FG (2003) Contaminação humana e ambiental por chumbo no Vale do Ribeira, nos Estados de São Paulo e Paraná, Brasil. Tese-Doutorado, Instituto de Geociências, Universidade Estadual de Campinas, Campinas.

Donni VL, Bagnoli P, Bartoli D, Bavazzano P, Cavalli IP, Landucci D, Marinari MG, Moggi A, Paoli L, Sannino G, Vannuchi C, Vitti A (1998) Blood lead levels in a non-professionally-exposed population from six Tuscan provinces. Ann Ist Sup Sanita 34(1):75–85.

Elhelu MA, Caldwell DT, Hirpassa WD (1995) Lead in inner-city soil and its possible contribution to children's blood lead. Arch Environ Health 50:165–169.

Eysink GGJ, Padua HB, Piva-Bertolettis AE, Martins MC, Pereira DN (1988) Metais pesados no Vale do Ribeira e Iguape-Cananéia. Ambiente: Rev CETESB de Tecnologia 2(1):6–13.

Fernícola NAGG, Azevedo FA (1981) Níveis de chumbo e atividade da desidratase do ácido δ-aminolevulínico (δ-ALA D) no sangue da população da grande São Paulo, Brasil. Rev Saúde Publica 15:272–282.

Flores J, Albert, L (2004) Environmental lead in México, 1990–2002. Rev Environ Contam Toxicol 181:37–109.

Freitas C, Simonetti MH, Silva MRP, Sakuma AM, Carvalho MFH, Duran MC, Kira CS, Tiglea P, Abreu MH, De Capitani EM (2002) Epidemiological investigation of children exposed to lead in the city of Bauru, São Paulo, Brazil. Epidemiology 14(5): 565–566.

Garrido NS, Pregnolato NP, Murata LTF, Silva MR, Nunes MCD, Engler VM, Sakuma AM (1990) Determinação de chumbo e cádmio em artigos escolares. Rev Inst Adolfo Lutz 50(1/2):291–296.

Garrido NS, Pregnolatto NP, Murata LTF, Silva MR, Nunes MCD, Antunes JLF, Tiglea P (1991) Avaliação dos níveis de arsênio, chumbo e cádmio em corantes e pigmentos utilizados em embalagens para alimentos no período de 1982 a 1989. Rev Inst Adolfo Lutz 51(1/2):63–68.

Gerhardsson L, Kazantzis G, Schultz A (1996) Evaluation of selected publications on reference values for lead in blood. Scand J Work Environ Health 22:325–331.

Jordão CP (1999) Distribution of heavy metals in environmental samples near smelters and mining areas in Brazil. Environ Technol 20(5):489–498.

Kuno R, Fernícola NAGG, Queiroz IR, Barros RCM (1993) O conteúdo de chumbo nos peixes consumidos pela população e os efeitos para a saúde. CETESB–Companhia de Tecnologia de Saneamento Ambiental, Diretoria de Normas e Padrões Ambientais. Departamento de Qualidade Ambiental e Padrões. Divisão de Padrões Ambientais e Toxicologia, Setor de Toxicologia Humana e Saúde Ambiental, São Paulo, Brazil.

Kuno R, Oliveira Filha MT, Sitnik RH (1994) Níveis de plumbemia de um grupo populacional próximo à Indústria Incometal–Pindorama–SP. CETESB–Companhia de Tecnologia de Saneamento Ambiental. Setor de Toxicologia Humana, São Paulo, Brazil.

Kuno R, Roqueti-Humayta MH, Oliveira Filha, M (1995) Níveis de plumbemia de um grupo populacional e animais de propriedades vizinhas à Indústria Tonolli S/A, Jacareí, São Paulo. Rev Bras Toxicol 8(1):152.

Kuno R, Fernícola NAGG, Campos AEM, Humaytá MHR, Oliveira Filha MT, Peres MS (1999) Níveis de chumbo no sangue de galináceos da região rural de Caçapava, S. P., Brasil. Rev Bras Toxicol 12(1):37–40.

Lanphear BP, Matte TD, Rogers, J, Clickner, RP, Dietz B, Bornschein RL, Succop P, Mahaffey KR (1998) The contribution of lead contaminated house dust and residential soil to children's blood lead levels. Environ Res 79:51–68.

Leroyer A, Nisse C, Hemon D, Gruchociak A, Salomez JL, Haguenoer JM (2000) Environmental lead exposure in a population of children in northern France: factors affecting lead burden. Am J Ind Med 38:281–289.

Lima MVDO, Pimentel MF, Brayner FMM, Abreu CAMA (2002) Lead released by handmade glazed ceramic food wares. Rev Bras Toxicol 15(1):19–23.

Liou SH, Wu TN, Chiang HC, Yang T, Wu YQ, Lai JS, Ho ST, Lee CC, Ko YC, Ko KN, Chabg PY (1996) Blood lead levels in Taiwanese adults: distribution and influencing factors. Sci Total Environ 180:211–219.

Machado LL (2001) Cádmio, chumbo e mercúrio em medicamentos fitoterápicos. Mestrado, Universidade de Brasília, Brasília.

Machado IC, Maio FD, Kira CS, Carvalho MFH (2002) Pb, Cd, Hg, Cu and Zn in mangrove oyster *Crassostrea brasiliana* Cananéia estuary, São Paulo, Brazil. Rev Inst Adolfo Lutz 61(1):13–18.

Maio FD, Okada IA, Carvalho MFH, Kira CS, Duran MC, Zenebon O (2002) Evaluation of the labelling, the minerals and inorganic contaminants in domestic and imported mineral water. Rev Inst Adolfo Lutz 61(1):27–32.

Marçal WS, Neto OC, Barreiros TRR, Hoshi EH, Moreno AM (2001a) Livestock as biomonitor to the environment impacts caused by industrial pollution in rural áreas. R Bras Med Vet 23(4):172–176.

Marçal WS, Gastel L, Liboni M, Pardo PE, Nascimento MR, Hisasi CS (2001b) Concentration of lead in mineral salt mixtures used as supplements in cattle food. Exp Toxicol Pathol 53:7–9.

Mazzeo J (1984) Ocorrência de saturnismo em eqüinos no Estado de São Paulo. Biológico 50(5):115–117.

Meyer I, Heinrich J, Trepka MJ, Krause C, Schultz C, Meyer E, Lippold U (1998) The effect of lead in tap water on blood lead in children in a smelter town. Sci Total Environ 209:255–271.

Meyer I, Heinrich J, Lippold U (1999) Factors affecting lead, cadmium, and arsenic levels in house dust in a smelter town in eastern Germany. Environ Res 81:32–44.

Minoia C, Sabbion E, Apostoli P, Pietra R, Pozzoli L, Gallorini M (1990) Trace element reference values in tissues from habitants of the European Community: a study of 46 elements in urine, blood and serum of Italian subjects. Sci Total Environ 95:89–105.

Montesanti M, Morgantini F, Landucci C, Rossi L, Biagi C, Castelli S (1995) Livelli ematici di piombo nei bambini della Pianna di Lucca. Rilevazione Del 1993. Minerva Pediatr 47:119–125.

Morisi G, Patriarca M, Carrieri MP, Fondi G, Tagi F (1989) Lead exposure: assessment of the risk for the general italian population. Ann Ist Sup Sanita 25:423–426.

Moura M (1996) A plumbemia na gravidez em um grupo de gestantes residentes na cidade do Rio de Janeiro, Brasil. Mestrado, Fundação Oswaldo Cruz, Saúde Pública.

Murgueytio AM, Evans RG, Sterling DA, Clardy SA, Shadel BN, Clements BW (1998) Relationship between lead mining and blood lead levels in children. Arch Environ Health 53:414–423.

NR7–Programa de Controle Médico de Saúde Ocupacional (2000) In: Segurança e Medicina do Trabalho, 45th ed. Atlas, São Paulo, pp 86–97.

Okada IA, Sakuma AM, Maio FD, Dovidauskas S, Zenebon O (1997) Evaluation of lead and cadmium levels in milk due to environmental contamination in the Paraíba Valley region of Southeastern Brazil. Rev Saúde Pública 31(2):140–143.

Paoliello MMB (2002) Exposição humana ao chumbo em áreas de mineração, Vale do Ribeira, Brasil. Tese-Doutorado, Faculdade de Ciências Médicas, Universidade Estadual de Campinas, Campinas.

Paoliello MMB, Chasin AAM (2001) Ecotoxicologia do chumbo e seus compostos. Série Cadernos de Referência Ambiental, vol 3. Cra-Centro de Recursos Ambientais, Salvador, Brasil.

Paoliello MMB, De Capitani EM (2003) Chumbo, 1st. In: Metais: gerenciamento da toxicidade. Atheneu, São Paulo, pp 353–398.

Paoliello MMB, Gutierrez PR, Turini CA, Matsuo T, Mezzaroba L, Barbosa DS, Carvalho SRQ, Alvarenga ALP, Rezende MI, Figueiroa GA, Leite VGM, Gutierrez AC, Lobo BCR, Cascales RA (2001) Reference values for lead in blood in an urban population in southern Brazil. Pan Am J Public Health 9(5):315–319.

Paoliello MMB, De Capitani EM, Cunha FG, Matsuo T, Carvalho MF, Sakuma A, Figueiredo BR (2002) Exposure of children to lead and cadmium from a mining area of Brazil. Environ Res 88:120–128.

Paoliello MMB, De Capitani EM, Cunha FG, Carvalho MF, Matsuo T, Sakuma A, Figueiredo BR (2003) Determinants of blood lead levels in an adult population from a mining area in Brazil. J Physiol 107:127–130.

Pereira OM, Henriques MB, Zenebon O, Sakuma A, Kira CS (2002) Determination of Pb, Cd, Hg, Cu and Zn levels in molluscs (*Crassostrea brasiliana, Perna perna* and *Mytella falcate*) Rev Inst Adolfo Lutz 61(1):19–25.

Pompéia ASL, Aguiar LSJ, Azevedo CMA, Eysink GG, Moraes RP, Hatamura E, Salomon VE, Cobra FB, Schmidt JR (1993) Contaminação ambiental por chumbo em torno da Indústria FAE S/A, Indústria e Comércio de Metais, Caçapava, SP. CETESB–Companhia de Tecnologia de Saneamento Ambiental. Informativo Técnico n° 004/93-DPT. Diretoria de Pesquisa e Desenvolvimento de Tecnologia. Departamento de Tecnologia de Proteção e Recuperação Ambiental. Divisão de Tecnologia de Recuperação Ambiental, Caçapava, Brazil.

Quitério SL, Silva CRS, Vaitsman DS, Martinhon PT, Moreira MFR, Araújo UC, Mattos, RCOC, Santos LSC (2001) Use of dust and air as indicators of environmental pollution in areas adjacent to a source of stationary lead emission. Cad Saúde Pública 17(3):501–508.

Romano J, Godinho R, Alonso CD, Martins MHRB (1992) Ethanol induced changes in the atmospheric lead in São Paulo Metropolitan Area, Brazil. In: Proceedings of the 9th Word Air Clean Congress, Montreal. CETESB–Companhia de Tecnologia de

Saneamento Ambiental, Diretoria de Normas e Padrões Ambientais, Departamento de Qualidade Ambiental e Padrões. Divisão de Qualidade do ar. Setor de Amostragem e Análise do Ar, São Paulo, Brazil.

Rocha AA, Pereira DN, Pádua HB (1985) Produtos de pesca e contaminates químicos na água da represa Billings, São Paulo, Brasil. Rev Saúde Publ 19:401–410.

Sakuma AM, Scorsafava MA, Zenebon O, Tiglea P, Fukumoto CJ (1989) Hortaliças comercializadas em São Paulo: aspectos da contaminação por chumbo, cádmio e zinco. Rev Inst Adolfo Lutz 49(1):81–84.

Santos Filho E, Silva RS, Barreto HHC, Inomata ONK, Lemes VRR, Sakuma AM, Scorsafava MA (1993) Concentrações sanguíneas de metais pesados e praguicidas organoclorados em crianças de 1 a 10 anos. Rev Saúde Pública 27(1):59–67.

Sepúlveda A, Veja J, Delgado I (2000) Exposición severa a plomo ambiental en una población infantil de Antofagasta, Chile. Rev Med Chile 128:221–232.

Silvany-Neto AM, Carvalho FM, Chaves MEC, Brandão AM, Tavares TM (1989) Repeated surveillance of lead poisoning among children. Sci Total Environ 78:179–186.

Silvany-Neto AM, Carvalho FM, Tavares TM, Guimarães GC, Amorin CJB, Peres MFT, Lopes RS, Rocha CM, Raña MC (1996) Lead poisoning among children of Santo Amaro, Bahia, Brazil in 1980, 1985, and 1992. Pan Am Health Organ Bull 30:51–62.

Spínola AG, Fernícola NAGG, Mendes R (1980) Intoxicação profissional por chumbo. In: Mendes R (ed) Medicina do Trabalho. Doenças Profissionais, 1st ed. Sarvier, São Paulo, pp 437–460.

Tavares TM, Brandão AM, Chaves MEC, Silany-Neto AM, Carvalho FM (1989) Lead in hair of children exposed to gross environmental pollution. J Environ Anal Chem 36:221–230.

Thornton I, Ramsey M, Atkison N (1995) Metals in the global environment: facts and misconceptions. ICME, Ontário, Canada.

Trepka MJ, Heinrich J, Krause C, Schultz C, Lippold U, Meyer E, Wichmann HE (1997) The internal burden of lead among children in a smelter town. Environ Res 72:118–130.

UNEP (United States Environmental Program) (1999) Phasing lead out of gasoline: an examination of policy approaches in different countries. http://www.unepie.org/energy/activities/Transport/Lead-gas/Lead.html. Accessed May 2, 2002.

WHO (1989) Environmental health criteria 85: lead-environmental aspects. WHO (World Health Organization), IPCS, Geneva. (Published under the joint sponsorship of the United nations Environment programme, the International Labour Organization, and the World Health Organization.)

WHO (1995) Environmental health criteria 165: inorganic lead. WHO (World Health Organization), IPCS, Geneva. (Published under the joint sponsorship of the United Nations Environment Programme, the International Labour Organization, and the World Health Organization.)

Yabe MJS, Oliveira E (1998) Metais pesados em águas superficiais como estratégia de caracterização de bacias hidrográficas. Química Nova 21(5):551–556.

Yang JS, Kang SK, Park IJ, Rhee KY, Moon YH, Sohn DH (1996) Lead concentration in blood among the general population of Korea. Int Arch Occup Environ Health 68:199–202.

Zhang ZW, Moon CS, Watanabe T, Shimbo S, He FS, Wu YQ, Zhou SF, Su DM, Qu IB, Ikeda M (1997) Background exposure of urban populations to lead and cadmium: comparison between China and Japan. Int Arch Occup Health 69:273–281.

Manuscript received March 25; accepted March 30, 2004.

Arsenic Speciation and Toxicity in Biological Systems

Kazi Farzana Akter, Gary Owens, David E. Davey, and Ravi Naidu

Contents

I. Introduction	97
II. Sources and Distribution of Arsenic	99
A. Natural or Geogenic Sources	99
B. Anthropogenic Sources	103
III. Chemistry of Arsenic	104
A. Groundwater Chemistry	105
B. Soil Chemistry	107
C. Soil and Water Cycles/Transfer Pathways	108
IV. Toxicity	109
A. Chemical Forms	109
B. Mechanisms of Toxicity	110
C. Toxic Effects on Humans and Animals	111
D. Toxic Effects on Plants	113
E. Exposure Pathway	115
V. Speciation	119
A. Definition of Speciation	119
B. Need for Speciation	120
C. Speciation in Water, Soil, and Crops	120
D. Methods of Speciation	121
E. Role of Speciation in Bioavailability	125
VI. Bioavailability	125
A. Definition of Bioavailability	125
B. Factors Affecting Bioavailability	131
C. Implications of Bioavailability to Toxicological Studies	133
VII. Conclusions	134
Summary	134
Acknowledgments	135
References	135

I. Introduction

The incidence of arsenic (As) contamination of groundwater systems in the Indian subcontinent, especially Bangladesh where thousands of people are dying of arsenicosis, has raised worldwide concern because millions of people are

---

Communicated by George W. Ware.

K.F. Akter · G. Owens · R. Naidu (✉)
Australian Centre for Environmental Risk Assessment and Remediation, University of South Australia, Mawson Lakes Campus, Mawson Lakes, SA 5095.

K.F. Akter · D.E. Davey
School of PMBS, University of South Australia, Mawson Lakes Campus, Mawson Lakes, SA 5095.

potentially at risk (Smith et al. 2003). Although As is potentially toxic to humans, animals, and plants, its actual toxicity varies widely with its oxidation state. Inorganic species are generally more toxic than organic species, and arsenite (As^{III}) is 60 times more toxic than arsenate (As^V), which is 70 times more toxic than methylated species, monomethylarsonic acid (MMA), and dimethylarsinic acid (DMA) (WHO 1981). The methylated species are consequently considered to be moderately toxic, whereas arsenobetaine (AsB) and arsenocholine (AsC) are considered to be nontoxic (Kumaresan and Riyazuddin 2001). Therefore, to provide an accurate view of environmental and human health risk assessment, identification and quantification of each species is essential, rather than just the total concentrations as has been the norm for many years. However, much of the published data concerned with risk assessment are generally based on laboratory studies, which may have only limited relevance to the complex heterogeneous systems commonly observed in the terrestrial environment and biological systems such as the human and animal gut. In addition, the number of speciation studies of As in environmental systems is scant, and hence there is a need to not only develop analytical speciation methods but to apply these methods to "real world" environmental matrices.

Arsenic toxicity is also directly related to the bioavailable fraction. Consequently, research on bioavailability is increasingly being recognized as critical for both environmental and human health risk assessment. Most bioavalability studies have been conducted on contaminated soil via the soil ingestion pathway (Rodriguez et al. 1999, 2003; Basta et al. 2001). Additionally, few researchers have investigated phytoavailability in contaminated soils. However, because phytoavailability varies with plant species and soil conditions, much research needs to be done to obtain realistic measures of As bioavailability.

Given the large-scale human poisoning via ingestion of As in many Asian countries, it is critical that for any risk-based assessment and remedial option, both the concentration and nature of the different As species in the terrestrial environment and biological systems should be assessed using tools capable of low-level detection. Simultaneously, research needs to examine As dynamics in the soil–plant system under field conditions that assess plant uptake of As to evaluate their potential impact on human and animal health. Several reviews have examined the sources and behavior of As in natural waters (Smedley and Kinniburgh 2002; Mukherjee and Bhattacharya 2001), its chemistry in soil (Smith et al. 1998), and the environmental chemistry of As (McNeill et al. 2001), but none of these has dealt specifically with speciation and toxicity in biological systems and the relationship to bioavailability. Therefore, this review addresses As speciation in environmental samples and common methods of determination, toxicity of As species in biological systems including symptoms and mechanism of acute and chronic poisoning and potential exposure pathways to the environment. Phytoavailability and phytotoxicity of As including hyperaccumulating plants are also reviewed with a view to highlighting exposure pathways and phytoremediation potential.

II. Sources and Distribution of Arsenic

Arsenic environmental inputs can be through either natural (geogenic) or anthropogenic processes. Natural processes including volcanic eruption, weathering of rocks and minerals, fossil fuel, and forest fire can release huge amounts into the environment that may be transported over long distances as suspended particulates through both water and air. Among the anthropogenic processes, industrial effluents contribute the highest source, accounting for much of the widespread contamination (Bhattacharya et al. 2002). The following sections briefly consider each of these two pathways of release into the environment. For more detailed information on sources of As, readers are directed to excellent reviews by Smedley and Kinniburgh (2002), Bhattacharya et al. (2002), Smith et al. (1998).

A. Natural or Geogenic Sources

Minerals Arsenic occurs naturally as a major constituent of more than 245 minerals (Woolson 1983a), including elemental As, arsenides, sulfides, oxides, arsenates, and arsenites. The greatest concentration of these minerals occurs in mineralized regions, including the United States, Canada, Australia, and some European countries, and are found in close association with transition metals as well as cadmium (Cd), lead (Pb), silver (Ag), gold (Au), antimony (Sb), phosphorus (P), and molybdenum (Mo). The most abundant As minerals are arsenopyrite (FeAsS), anargite (Cu_3AsS_4), and realgar (As_4S_4) (Cullen and Reimer 1989; Woolson 1983a; NRC 1977). A selection of arsenic-bearing minerals and their average As concentration is shown in Table 1.

Although not a major component, As is also present to various degrees in

Table 1. Common arsenic-bearing minerals and ore associations.

Mineral	Formula	Occurrence in nature	%As (average)
Arsenopyrite	FeAsS	Lode gold (e.g., Bendigo and Stawell in Victoria, Australia)	0.5
		Cu sulfide	4.0
		Sn (e.g., Renison in Tasmania, Australia)	0.2
Cobaltite	CoAsS	Ni, Cu, and Zn ores	2.5
Enargite	Cu_3AsS_4	Hydrothermal vein and replacement (e.g., USA, Chile, Philippines)	0.1
Niccolite	NiAs	Vein deposits and norites (e.g., Kambalda in Western Australia)	0.5
Orpiment	As_2S_3	Hydrothermal veins	2.0

Source: Shraim 1999.

other common minerals, primarily sulfides such as pyrite. The As content of rock strata depends on the rock type, with sedimentary rocks containing much higher concentrations than igneous rock strata (Bhumbla and Keefer 1994). Concentration in sedimentary rocks is typically in the range of 5–10 mg/kg (Webster 1999). A summary of background concentrations in different minerals is presented in Table 2. Concentration varies between 0.6 and 120 mg/kg in sand and sandstones and may be as high as 490 mg/kg in shales and clay formations (Bhattacharya et al. 2002). Concentrations in igneous rocks are generally low, and there is little variation with igneous rock type. Ure and Berrow (1982) quoted an overall average value of 1.5 mg/kg for all igneous rock types.

Soils and Sediments The main source of As in soils is the parent rock from which the soil was derived (Yan-Chu 1994). Baseline concentrations in soils are generally of the order of 5–10 mg/kg (Smedley and Kinniburgh 2002) but may vary with the geological history of the region. Cullen and Reimer (1989) examined a wide range of soils and reported an average of 5–6 mg/kg for uncontaminated soils. The lowest concentration was found in sandy soils and those derived from granites, whereas larger concentrations were found in alluvial and organic soils (Mandal and Suzuki 2002). Acid sulfate soils that are generated by the oxidation of pyrite in sulfide-rich terrains such as pyritic shales, mineral veins, and dewatered mangrove swamps can also be relatively enriched. Dudas (1984) found concentration up to 45 mg/kg in the B horizons of acid sulfate soils derived from the weathering of pyrite-rich shales in Canada. The level of soil As varies widely with the country of origin (Table 3). This difference may be attributed to the widely different geology and minerals constituting the parent rocks. The natural level in sediments is usually below 10 mg/kg dry weight (Mandal and Suzuki 2002) and varies considerably all over the world. The high-

Table 2. Abundance of arsenic in crustal materials.

Rock type	Range (mg As/kg)
Igneous rocks	
Ultrabasics	0.3–16
Basalts	0.06–113
Andesites	0.5–5.8
Granites/silicic volcanics	0.2–13.8
Sedimentary rocks	
Shales and clays	0.3–490
Phosphorites	0.4–188
Sandstones	0.6–120
Limestones	0.1–20
Coals	0.5–80

Source: Reprinted with permission from Bhattacharya et al. 2002, © 2002 Marcel Dekker, Inc.

Table 3. Arsenic content in the soils of various countries.

Country	Type of soil sediment	Range (mg/kg)	References
West Bengal, India	Sediments	10–196	Chakraborti et al. 2001b
Bangladesh	Sediments	9–28	Nickson et al. 2000
Argentina	All types	0.8–22	Reichert and Trelles 1921 cited in Mandal and Suzuki 2002
China	All types	0.01–626	Wei et al. 1991
Italy	All types	1.8–60	Zuccuri 1913 cited in Mandal and Suzuki 2002
Japan	All types	0.4–70	Zuccuri 1913 cited in Mandal and Suzuki 2002
Mexico	All types	2–40	Whetstone et al. 1942
United States	All types	1–20	Stater et al. 1937

Source: Adapted from Mandal and Suzuki 2002.

est concentration, 196 mg/kg, was reported for a Gangetic sediment (Chakraborti et al. 2001b).

Water Arsenic is typically found in relatively low concentrations in natural waters with a concentration dependent on the local geology, hydrology, and geochemical characteristics of the aquifer materials (Bhattacharya et al. 1997). Weathering and leaching of arsenic-rich geological formations and mining wastes result in elevated concentrations in natural waters worldwide. The risk of As contamination is, however, much higher in groundwater than in surface waters such as rivers, lakes, and reservoirs. The lower concentrations in rivers and lakes are probably due to aerobic oxidation of As and subsequent attenuation by oxidic minerals as well as surface recharge and runoff.

The USEPA maximum permissible concentration of As in drinking water is 20 μg/L (USEPA, revised 2001) and WHO (1998) recommended value is 10 μg/L. Incidences of the natural occurrence in groundwater (>10 μg/L) are reported in several countries (Table 4).

Arsenic is also the 10th most abundant element in seawater, with an average value of about 2 μg/L (Cullen and Reimer 1989; Russeva and Harvezo 1993). In comparison, rainwater derived from an uncontaminated mass of oceanic air contains an average 19 ng/L (Andreae 1980). In natural lakes, levels range from 0.2 to 56 μg/L (Crecelius 1975). River waters generally contain low concentrations, but these can be elevated due to the presence of As in suspended particulates and from anthropogenic sources. Concentration in unpolluted freshwaters typically ranges from 1 to 10 μg/L, increasing to 100–5000 μg/L in areas of sulfide mineralization and mining (Smedley et al. 1996). The highest levels recorded were in waters from areas of thermal activity in New Zealand (Mandal

Table 4. Concentrations of arsenic in groundwater of arsenic-affected countries.

Country/region	Arsenic source	As concentration (µg/L)	References
Bangladesh	Well water	0.5–2500	BGS/DPHE 2001; DPHE/BGS/MML 1999; Karim 2000
West Bengal, India	Well water	<1–1300	Bhattacharya et al. 1997; DPHE/BGS/MML 1999; Mandal et al. 1996
China, Inner Mongolia	Well water	<100–1860	Luo et al. 1997; Lianfang and Jianzhong 1994
Taiwan	Deep well water	Up to 1800	Chen et al. 1994
Thailand	Shallow well water	120–6700	Choprapawon and Rodcline 1997
Ghana	Well water	2–175	Smedley 1996
Argentina	Shallow aquifers	100–4800	Smedley et al. 1998; Bundschuh et al. 2000
Chile	Well water	100–1000	Borgono et al. 1977
Mexico	Well water	300–1100	Armienta et al. 1997, 2000
Hungary	Deep well water	25–>50	Varsanayi et al. 1991
U.S.A.	Well water	100–>500	Welch et al. 2000
Canada	Well water	18–146	Grantham and Jones 1977
Vietnam	Arsenic-rich sediment	1–3050	Berg et al. 2001

Source: Adapted from Smedley and Kinniburgh 2002; Mukherjee and Bhattacharya 2001.

and Suzuki 2002), up to 8.5 mg/L, and in geothermal waters in Japan (Mandal and Suzuki 2002), 1.8–6.4 mg/L.

Air The concentration of As in the atmosphere is usually low and exists predominantly as adsorbed particulate matter. Concentration in air typically ranges from 10^{-5} to 10^{-3} µg/m^{-3} in unpolluted areas, increasing to 0.003–0.18 µg/m^{-3} in urban areas and greater than 1 µg/m^{-3} close to industrial areas (WHO 1998). Typical levels for the European region are currently quoted as being from 0.2 to 1.5 ng/m^{-3} in rural areas, 0.5–3 ng/m^{-3} in urban areas, and no more than 50 ng/m^{-3} in industrial areas (DG Environment 2000). About 60% of the atmospheric flux has been estimated to be caused by low-temperature volatilization, with volcanic activity being the next most important natural source (Chilvers and Peterson 1987). Emission through volcanic eruption is mostly in the form of dust, about 0.3 Gg/yr, compared to 0.01 Gg/yr in volatile forms (Bhattacharya et al. 2002).

B. Anthropogenic Sources

Anthropogenic activities are the main source of As in the environment, exceeding natural sources by 3:1 (Woolson 1983a). Among the anthropogenic sources, industrial effluents constitute the largest contribution. Approximately 30,000 industrial facilities exist in and around major cities where soil As concentration ranged from 2.13 to 4.27 mg/kg (Mukherjee and Bhattacharya 2001). In developed and developing regions, industrial sources generally include coal-fired power plants, smelting, incinerations of wastes, wood preservation, and agriculture. About 90% of industrial As in the United States is used for wood preservation, but As is also commonly used in the manufacture of paints, dyes, ceramics and glass, electronics, pigments, and antifouling agents (Leonard 1991). The following sections briefly consider the sources and forms entering the environment through human activities.

Industry Arsenic trioxide (As_2O_3) is the major form of As produced for industry by the U.S., Sweden, France, the former USSR, Mexico, and southwest Africa. As_2O_3 is used to decolorize glass and in the manufacture of pharmaceuticals. It can be recovered from smelting or roasting of nonferrous metal ores. In 1990, the world production of As_2O_3 reached its peak at 50,000 t/yr and has since declined (Loebenstein 1993). Nriagu and Pacyna (1988) estimated 64,000–132,000 t/yr total worldwide anthropogenic As discharge onto land, of which commercial wastes contribute about 40%, coal ash about 22%, mining industry about 16%, and atmospheric fallout from the production of steel about 13%.

Smelting and Mining Elevated concentrations of As, as well as other metals such as Cd, Cu, Fe, Pb, Ni, and Zn, are commonly encountered in acid mine effluents. Consequently, contamination of the atmosphere, soils, sediments, streams, and groundwater is possible during mining and/or smelting. In a goldfield in Ballarat, Victoria, Australia, Lamb et al. (1996) reported 1000–2650 mg/kg As in quartz sand of the gold field. Crecelius et al. (1974) reported that a large Cu mine near Tacoma, WA, released approximately 300 t of particulate As matter into the atmosphere annually. Tailings from metal sulfide mining operations release substantial amounts of As and heavy metals into groundwater and surface water because of the oxidation of arsenopyrite (FeAsS).

Agricultural Chemicals Agricultural inputs from chemicals such as insecticides, herbicides, desiccants, and fertilizers are a major source of As in soils (Jiang and Singh 1994). Over the past 100 years, approximately 80% of the As produced was used in the manufacture of insecticides, herbicides, and fungicides. Before the introduction of dichlorodiphenyltrichloroethane (DDT), Pb arsenate (PbAsO$_4$), calcium arsenate (CaAsO$_4$), magnesium arsenate (MgAsO$_4$), zinc arsenate (ZnAsO$_4$), and Paris green (copper acetoarsenite) were used extensively as insecticides in orchards (Merry et al. 1983). Arsenic acid (H$_3$AsO$_4$), monosodium methanearsonate (MSMA), disodium methanearsonate (DSMA), and cacodylic acid (CA) are still used as herbicides in cotton production.

Arsenical pesticides were also widely used in livestock dips to control ticks, fleas, and lice (Vaughan 1993). Sodium arsenite was extensively used in cattle dips in all states of Australia, but particularly in Queensland and New South Wales (NSW), from the turn of the 20th century until the early 1950s. This practice has resulted in thousands of contaminated dip sites distributed throughout Australia (Chappell et al. 1995; McLaren et al. 1996; McDougall 1996), and more than 1000 cattle dips are still in operation in NSW (McDougall 1996). Moreover, inorganic arsenical compounds were used as nonselective soil sterilants and herbicides until the late 19th century.

Wood Preservatives The use of chromated copper arsenate (CCA) and other As-based chemicals in wood preservation industries has caused widespread contamination of soils and the aquatic environment (Jacks and Bhattacharya 1998; Bhattacharya et al. 1995). CCA had attained wide-scale industrial application as a wood preservative owing to the biocidic characteristics of Cu^{II} and As^{V}. Earlier, CCA and ammoniacal copper arsenate (ACA) were used in combination in 90% of the arsenical wood preservatives (Perker 1981). Studies around an abandoned wood preservation site at Konsterud, Kristinehamns community in Central Sweden (Bhattacharya et al. 1995, 1996) revealed soil As concentrations between 10 and 1067 mg/kg, and the order of abundance for metal contaminants was found to be As > Zn > Cu > Cr.

III. Chemistry of Arsenic

Arsenic was first isolated by Albertus Magnus in 1250 CE. Although As was officially described as a group VA nonmetal in the periodic table (Nelson 1983), it is commonly referred to as a metalloid because it shares similar physical and chemical properties with metals and nonmetals (Ringwood 1995; Clifford and Zhang 1993; WHO 1981).

Arsenic can be found in both organic and inorganic compounds with variable oxidation states. The common oxidation states are arsenate (As^{V}), arsenite (As^{III}), elemental arsenic (As^{0}), and arsine (As^{-III}) (Clifford and Zhang 1993; Cullen and Reimer 1989; Ferguson and Gavis 1972; Amran et al. 1995; Haswell et al. 1985; Hung and Liao 1996). Arsenic can also exist as oxyanions such as AsO$_4^{3-}$ and AsO$_3^{3-}$. Species having the lower oxidation states (0 and −3) occur only under strongly reducing conditions, whereas species of the higher oxidation states (+5 and +3) occur under oxygenated and mildly reducing conditions, respectively (Hung and Liao 1996). The chemical structures of As species that are environmentally important are shown in Fig. 1.

Chemically, As is very similar to its neighbor phosphorus in the periodic table (Reuther 1992; NRC 1977; Clifford and Zhang 1993; Cullen and Reimer 1989). Thus, the behavior of arsenate and *o*-phosphate are very similar, and biological uptake of arsenate is claimed to be dependent on the concentration of *o*-phosphate due to competition for absorption sites (Reuther 1992; Cullen and Reimer 1989).

A. Groundwater Chemistry

The origin and mobility of As in the groundwater environment have received significant attention in recent years. Elevated concentrations in natural waters are known to have resulted from weathering and leaching of As-rich geological formations, drainage from mine tailings and wastes, and from thermal springs.

In groundwater, inorganic As commonly exists as either arsenate (As^V) or arsenite (As^{III}), the latter being considered to be more mobile and more toxic to

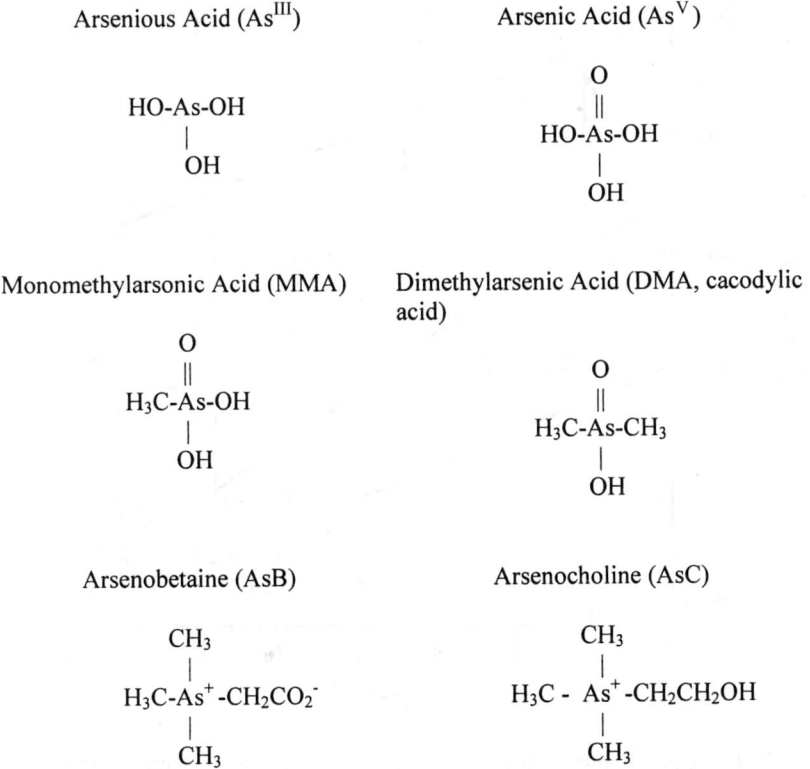

Fig. 1. Chemical structures of some toxicologically relevant arsenic species.

living organisms (Korte and Fernando 1991). In aqueous environments, prokaryotes and eukaryotes reductively biomethylate inorganic As to dimethylarsinic acid (DMAA) and monomethylarsonic acid (MMAA) (Bhattacharya et al. 2002), but the toxicity of these methylated forms is less. Biomethylation affects the mobility and transport of As in groundwater (Maeda 1994). The pentavalent species are predominant and stable in the oxygen-rich aerobic environments, whereas the trivalent species are predominant in a moderately reducing environment. Kim et al. (2002) studied groundwater samples from southeast Michigan and reported that most (53%–98%) of the detected As species was arsenite (As^{III}). Redox potential (Eh), pH, and dissolved oxygen (DO) are all important factors controlling As speciation and chemistry in groundwater. Under oxidizing conditions at pH <2, As^V occurs as H_3AsO_4 in the Eh–pH diagram (Fig. 2; Smedley and Kinniburgh 2002). In the pH range from 2 to 14, H_3AsO_4 dissociates to $H_2AsO_4^-$, $HAsO_4^{2-}$, and AsO_4^{3-}. At low Eh values, As becomes dominant as $H_3AsO_3^0$. Up to pH 9, H_3AsO_3 does not dissociate; above pH 9, it appears as $H_2AsO_3^-$, $HAsO_3^{2-}$, AsO_3^{3-}.

Sadiq and Alam (1996) studied As chemistry in a groundwater aquifer from

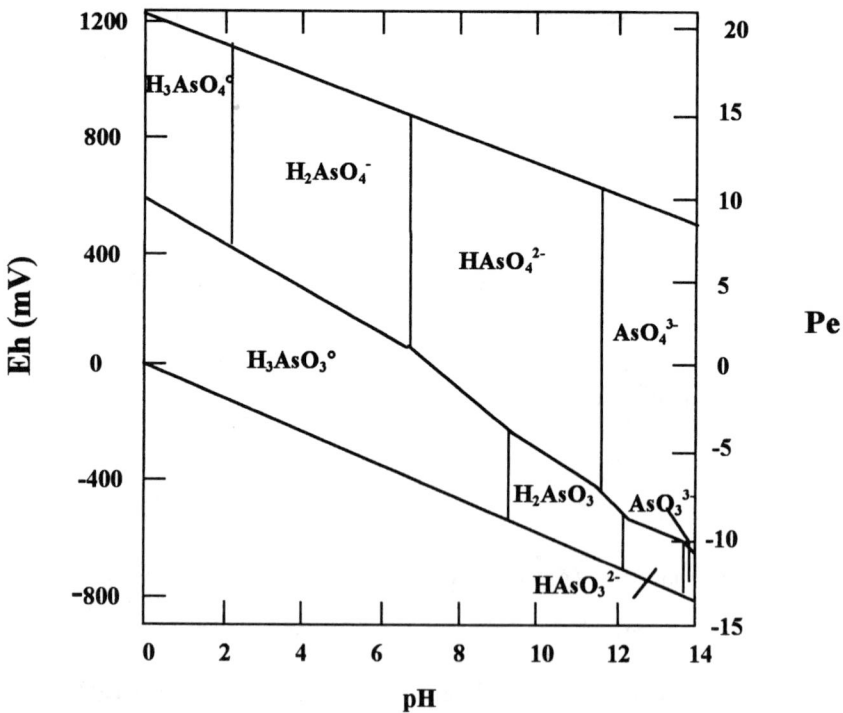

Fig. 2. Redox potential (Eh)–pH diagram for aqueous arsenic species in the system As-O_2-H_2O at 25°C and 1 bar total pressure. (Reprinted with permission from Smedley and Kinniburgh 2002; © 2002 Elsevier science).

the eastern province of Saudi Arabia and found that $H_2AsO_4^-$ was the dominant species in acidic groundwater and $HAsO_4^{2-}$ the most abundant species in alkaline groundwater. In the presence of extremely high concentrations of reduced sulfur, dissolved arsenic–sulfide species can be significant. Reducing acidic conditions favor precipitation of orpiment (As_2S_3), realgar (AsS), or other sulfide minerals containing coprecipitated As. Therefore, high levels of As in waters are not expected where there is a high concentration of free sulfide (Moore et al. 1988).

B. Soil Chemistry

In soils, as in water, the chemical behavior of As is in many ways similar to that of phosphoros (P), especially in aerated systems, where As^V ions generally resemble the orthophosphate ion closely (Walsh et al. 1977). However, under conditions normally encountered in soils, As is more mobile than P and unlike P can undergo changes in its valency (Smith et al. 1998).

Generally, As does not follow the typical behaviour of other metal contaminants. For instance, it is highly soluble in neutral to alkaline pH (6.6–7.8) whereas most of the heavy metals are more mobile under acidic conditions. However, As can also be moderately soluble under acidic conditions. For this reason, its chemistry is more complex in soils than many other pollutants. Arsenic forms a variety of inorganic and organic compounds in soils (Vaughan 1993) and is present as inorganic species, either As^V or As^{III} (Masscheleyn et al. 1991). The forms present in soils depend on the type and amounts of sorbing components of the soil, pH, and the redox potential (Yan-Chu 1994). Under oxic soil conditions (Eh > 200 mV; pH 5–8), As is commonly present in the +5 oxidation state. However, As^{III} is the predominant form under reducing conditions (Masscheleyn et al. 1991; Marin et al. 1993). Both As^V and As^{III} species can undergo chemical and or microbial oxidation-reduction and methylation reactions in soils and sediments and can adsorb on hydrous oxides of Fe, Al, and Mn (Bhumbla and Keefer 1994; Fig. 3.).The most important natural attenuation process known for As^{III} compounds is precipitation as As sulfide (As_2S_3).

As^{III} is more toxic and mobile in soils than As^V, and methylated species such as monomethylarsonic acid [MMAA, $CH_3AsO(OH)_2$] and dimethylarsinic acid [DMAA, $(CH_3)_2AsO(OH)$] are also mobile (Bhattacharya et al. 2002). However, these methylated forms are volatile and unstable under oxidizing conditions and are cycled back into the soil environment in inorganic forms (Cullen and Reimer 1989).

Ferric hydroxide [$Fe(OH)_3$] plays an important role in controlling the As concentration in soils. Both As^V and As^{III} are adsorbed onto the surface of $Fe(OH)_3$, but the adsorption of As^V is much higher than As^{III} (Bhattacharya et al. 2002). In general, highly oxidic soils sorb three times more As^V than soils containing small amounts of oxidic minerals (Smith et al. 1999). The presence of iron, aluminium, and calcium compounds is the most important factor in controlling the fixation in soil (Bissen and Frimmel 2003). The sorptive capacity of a soil for an ion is a function of its surface area and hence its clay content;

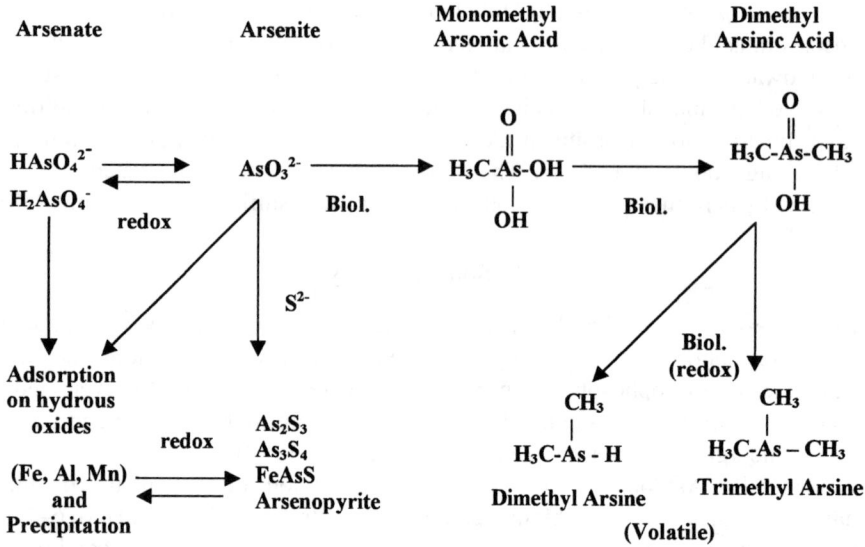

Fig. 3. Transformations of arsenic in soil. (Adapted from Bhumbla and Keefer 1994.)

this explains why As is more mobile in sandy soil (low clay content) than fine-textured soils (Walsh et al. 1977).

C. Soil and Water Cycles / Transfer Pathways

The principal natural reservoirs of As are rocks. Release and mobilization in various forms from these sources governs the availability of this element to soil and water. It is likely that the major mechanism of As release is through the decomposition of arsenopyrite, i.e., pyrite oxidation according to the following equation (Rimstidt et al. 1994):

$$FeAsS + 13Fe^{3+} + 8H_2O = 14Fe^{2+} + SO_4^{2-} + 13H^+ + H_3AsO_4 \text{ (aq)}$$

This reaction suggests that significant amounts of Fe^{2+} and SO_4^{2-} would be released into water together with As at low pH, which is contrary to the actual mechanism observed in some highly As-affected groundwaters. Mobilization in such groundwaters is primarily due to the desorption of As oxyanions from the surface of Fe-oxyhydroxides as well as reductive dissolution of the Fe-oxyhydroxides in the sediments (BGS MML 1999; Nickson et al. 1998). This process is driven by microbial oxidation of sedimentary organic matter by consuming dissolved oxygen and other oxidants from the aquifers, thereby increasing the groundwater alkalinity (Nickson et al. 2000; Mukherjee and Bhattacharya 2001). High alkaline groundwater in the reducing confined aquifers may be produced according to the reaction:

$$4FeOOH + CH_2O + 7H_2CO_3 = 4Fe^{2+} + 8HCO_3^- + 6 H_2O$$

This reductive dissolution of the Fe-oxyhydroxide release both Fe and As in aqueous solution and increase pH by releasing HCO_3^- as well.

Mukherjee and Bhattacharya (2001) suggested that the majority of As detected in sediments was quantitatively related to the amounts of the amorphous Fe-, Al-, and Mn-oxyhydroxides in those sediments. These surface reactive compounds are characterized by a positive charge and therefore act as a strong adsorbent for the As oxyanion, especially As^V (Pierce and Moore 1982). With the increase of pH (alkalinity), the surface reactive compounds attain zero charge and release the As oxyanions. The overall transfer pathway of arsenic is shown in Fig. 4.

IV. Toxicity
A. Chemical Forms

Arsenic is potentially toxic to humans, animals, and plants (Ringwood 1995; Gustafsson and Jacks 1995; Marin et al. 1993; Abernathy 1983). The toxicity of As depends on a number of factors, most importantly the chemical form (inorganic or organic) and the oxidation state of the arsenical. It is generally recognized that the soluble inorganic arsenicals are more toxic than the organics and that inorganic As^{III} (iAs^{III}) is more toxic than inorganic As^V (iAs^V) (UN 2001; Maeda 1994). The methylated metabolites such as monomethylarsinic acid (MMA) and dimethylarsenic acid (DMA) are considered to be nontoxic because methylation appears to be primarily a detoxification mechanism (Roy and Saha 2002). Recent studies suggest that methylation is not fully a detoxication process because it needs the methyltransferase enzyme to occur, and some mammalian species are deficient in this enzyme. According to Lin et al. (1999),

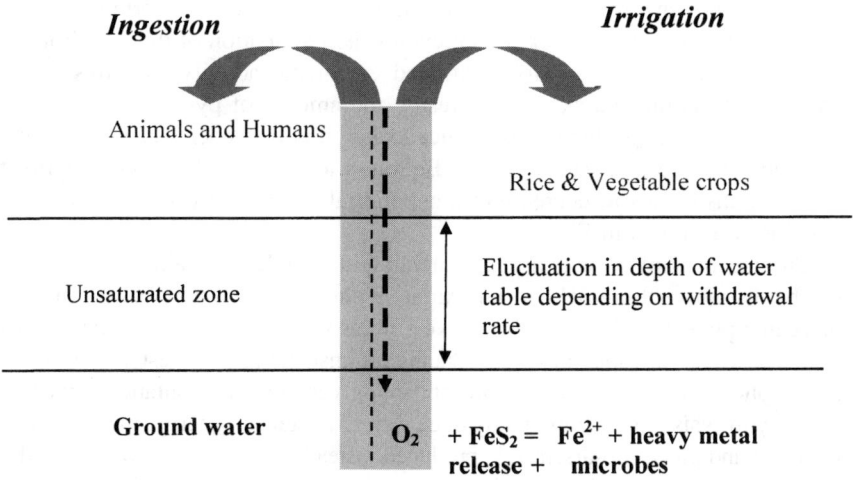

Fig. 4. Schematic representation of the arsenic transfer pathway in the environment.

MMA^{III} is more than 100 times more potent than inorganic As^{III} as an *in vitro* inhibitor of thioredoxin reductase. Thus, formation of MMA^{III} appears to be indicative of toxification of both inorganic As^{III} and inorganic As^{V}. Organoarsenicals such as AsB, AsC, and Arsenosugar are considered to be nontoxic and are common in many seafoods (Kumaresan and Riyazuddin 2001). Indeed, some forms, such as arsanilic acid, are not only nontoxic but are fed to poultry to improve feed efficiency, feathering, egg production, and pigmentation and to swine to control dysentery (NRC 1977). Vega et al. (2001) reported the toxicity order of arsenicals as iAs^{III} > monomethylarsine oxide (MMAO^{III}) > DMA^{III} > DMA^{V} > MMA^{V} > iAs^{V}. Previously, arsine gas (AsH$_3$) was considered the most toxic As compound due to its ability to break down the red blood cells in humans (Whitecare and Pearse 1974). However, it is unstable at high temperature and rapidly decomposes in the presence of light and moisture (Fowler 1976).

B. Mechanisms of Toxicity

The most common toxic mode of action for any element is the inactivation of enzyme systems, which serve as biological catalysts (Dhar et al. 1997). Arsenic causes inactivation of up to 200 enzymes, most notably those involved in the cellular energy pathway and DNA synthesis and repair. Inorganic As^{V} does not react directly with the active sites of enzymes (Johnstone 1963), and iAs^{V} is first reduced to iAs^{III} *in vivo* before a toxic effect is observed (Mandal and Suzuki 2002). Trivalent As interferes with enzymes by bonding to −SH and −OH groups, especially when there are two adjacent SH(−SH) groups in the enzyme. Cellular energy generated by enzymes in the citric acid cycle is adversely affected by iAs^{III}. iAs^{III} complexes with pyruvate dehydrogenase (PDH), inactivating the enzyme, and thereby the generation of adenosine-5-triposphate (ATP) is prevented. Inorganic As^{III} can replace two hydrogen atoms from the thiol groups of enzymes and complex with a sulfur molecule to form a dihydrolipoylarsenite chelate complex; this prevents the reoxidation of the dihydrolipoyl group, which is necessary for continued enzymatic activity, and this pivotal enzyme step is thus blocked. As a result, the amount of pyruvate in the blood increases, energy production is reduced, and finally the cell is progressively damaged (Mandal and Suzuki 2002; Belton et al. 1985). Inhibition of pyruvate oxidation also leads to decreased mitochondrial respiration (Fowler 1977). The reactions are shown in Fig. 5.

The mechanism by which pentavalent As acts is less certain. Inorganic As^{V} can be disruptive to mitochondrial oxidative phosphorylation by competing with inorganic phosphate because they have similar structures and properties (Dixon 1997). Arsenate disrupts this by producing an unstable arsenate ester of 1-arseno-3-phosphoglycerate instead of 1,3-diphosphoglycerate that spontaneously undergoes hydrolysis nonenzymatically (see Fig. 5). Hence, energy metabolism is inhibited and glucose-6-arsenate is produced instead of glucose-6-phosphate (Mandal and Suzuki 2002). Arsenate is also known to inhibit normal DNA repair and synthesis by replacing phosphoros in DNA (Fowler 1977). Most recently, An-

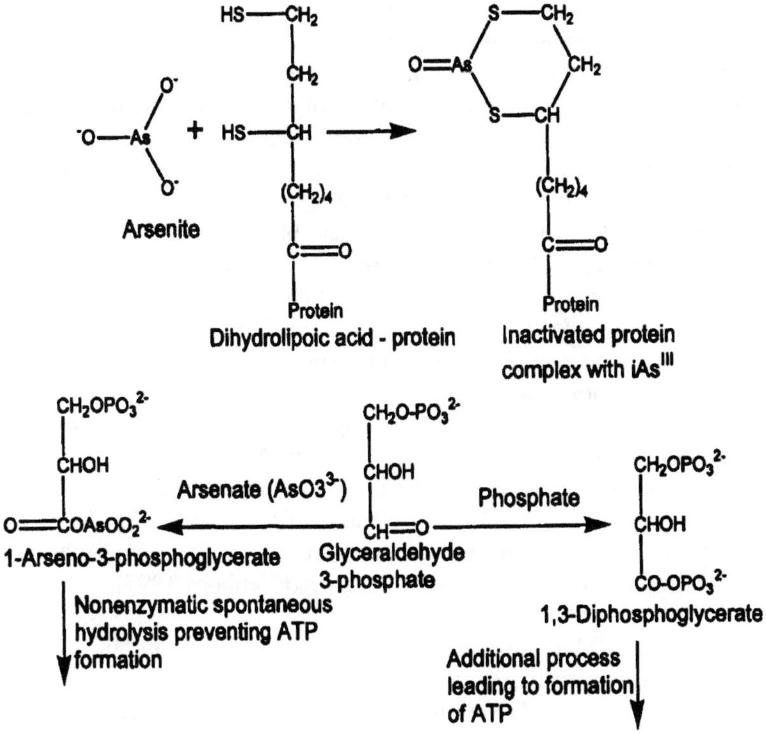

Fig. 5. Schematic representation of the inactivation of protein with As^{III} and the inhibition of ATP formation by replacing As^V instead of phosphate. (Reprinted with permission from Mandal et al. 2002; © 2002 Elsevier Science.)

drew et al. (2003) reported that As^{III} can reduce nucleotide excision repair capacity of damaged DNA by inhibiting expression critical genes involved in this function; however, this hypothesis needs further verification.

Methylated trivalent arsenicals such as MMA^{III} are potent inhibitors of glutathione (GSH) reductase (Styblo et al. 1997) and thioredoxin reductase (Lin et al. 1999). The inhibition may result from the interaction of trivalent As with critical thiol groups in these molecules. The activity of the methylated trivalent arsenicals is greater than arsenite, MMA^V, and DMA^V. Inhibition of these enzymes may alter cellular redox status and eventually lead to cytotoxicity.

C. Toxic Effects on Humans and Animals

For human and animal receptors, toxicity can be described either as being acute or chronic. Acute toxicity causes immediate adverse effects resulting from short-term exposure to high concentrations of As whereas chronic toxicity causes adverse effects resulting from long-term exposure to low levels. These two types of toxicity are briefly discussed.

Acute Toxicity The acute toxicity of As is related to its chemical form and oxidation state. Acute toxicity requires prompt medical attention and usually occurs through ingestion of contaminated food and drink. Symptoms of acute toxicity include gastrointestinal discomfort, diarrhea, and anemia have been reported in many countries, including Argentina, Chile, Taiwan, West Bengal (India), and Bangladesh (Guha Mazumder et al. 1992). The characteristics of severe acute toxicity in humans include burning and dryness of mouth and throat, gastrointestinal discomfort, vomiting, diarrhea, bloody urine, shock, convulsions, coma, and death. The LD_{50} values of several arsenicals in laboratory animals are displayed in Table 5, which shows significant variation among test species. Moreover, for the same species widely different LD_{50}s have been reported with variation in the nature of the compound. This difference may be attributed to differences in the bioavailable As in the compounds. Thus, those chemicals that are highly bioavailable will give a low LD_{50} value and pose higher risk. A basic tenet is that acute toxicity of As^{III} is greater than As^V. For example, in mice, the oral LD_{50} of arsenic trioxide (As_2O_3) is more than 20 fold lower than that of calcium arsenate. However, Petrick et al. (2001) reported that MMA^{III} has a lower LD_{50} than As^{III} in hamsters. In the human adult, the lethal range of inorganic As is estimated at a dose of 1–3 mg/kg (Ellenhorn 1997).

Chronic Toxicity Many different systems within the human body are affected by chronic exposure to inorganic As. Some of these systems are skin, respiratory, gastrointestinal, cardiovascular, nervous, hepatic, endocrine, and hematopoietic (Chen and Lin 1994; Chen et al. 1997). One of the hallmarks of chronic toxicity in humans indicative of oral exposure is skin lesions, which are characterized by hyperpigmentation, hyperkeratosis, and hypopigmentation (Cebrian et al. 1983). In Taiwan, Blackfoot disease, a peripheral vascular disease that

Table 5. Acute toxicity of arsenic in laboratory animals.

Arsenical	Species (sex)	Route	LD_{50} (mg As/kg)	Reference
Arsine	Rat	Oral	3	Amran et al. 1995
Sodium arsenite	Mouse	Oral	15–22	Donohue and Abernathy 1999
Arsenic trioxide	Mouse	Oral	34	Donohue and Abernathy 1999
Calcium arsenate	Mouse	Oral	20–800	Donohue and Abernathy 1999
MMA	Mouse	Oral	700–1800	Donohue and Abernathy 1999
DMA	Mouse	Oral	1200–2699	Donohue and Abernathy 1999
Arsenite	Hamster (m)	ip	8	Petrick et al. 2001
MMA^{III}	Hamster (m)	ip	2	Petrick et al. 2001
Arsenobetaine	Mouse	Oral	$\geq 10 \times 10^3$	Donohue and Abernathy 1999
Arsenocholine	Mouse	Oral	6.5×10^3	Donohue and Abernathy 1999

MMA, monomethylarsonic acid; DMA, dimethylarsinic acid; m, male; ip, intraperitoneal; LD, lethal dose.

leads to gangrene of the extremities, is also observed in individuals chronically exposed to As in their drinking water (Tseng 1977). Internal cancers of the lung, liver, bladder, and kidney have also been associated with chronic ingestion of As-contaminated drinking water (Chiou et al. 1995; Cuzick et al. 1992). Roy and Saha (2002) reported that inorganic arsenicals have the potential to cause skin cancers in humans if they are ingested for a long period. However, the exposure time required to develop arsenicosis varies from case to case, probably reflecting a dependence on As levels in drinking water and food, nutritional status, genetic variation in the human being, and other compounding factors. The overall toxic effect of As in humans is shown in Fig. 6.

As a result of the harmful effects of As on humans, animals, and plants, world regulatory organizations have established maximum allowable limits in drinking water, food, and soil (Table 6).

D. Toxic Effects on Plants

There is a paucity of information on the toxic effects of As on plants in terrestrial ecosystems. Many of the studies reported are based on hydroponic systems. Outcomes from such studies suggest that plants exposed to high levels from

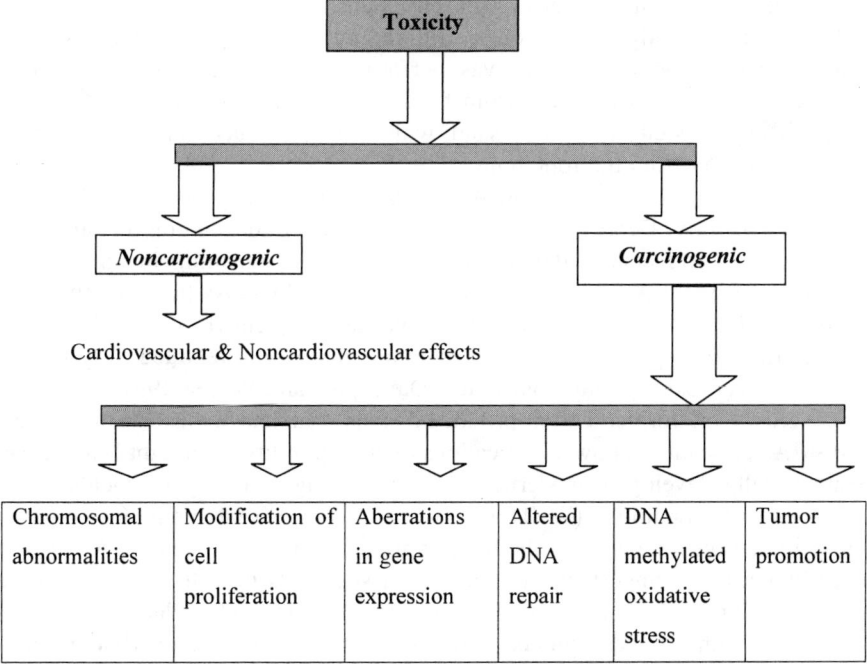

Fig. 6. Flowchart describing the various modes of arsenic toxicity. (Adapted from Roy and Saha 2002.)

Table 6. Recommended arsenic concentration guideline for drinking water, food, and soil.

Subjects	Items	WHO[a] 1996	ANZECC[b] 1995	USEPA[c] 2001
Water (mg/L)	Drinking	0.01	0.007	0.01
	Irrigation	—	0.1	—
Soil quality (mg/kg dry weight)	Environmental investigation criterion	—	20	—
	Human health investigation criteria	—	100	—
Food crops (mg/kg fresh weight)	Vegetables and fruits	2	1	0.5–1

[a]World Health Organization; [b]Australian and New Zealand Environmental Conservation Council; [c]U.S. Environmental Protection Agency.
Source: Shraim 1999.

irrigation water, arsenical insecticides and herbicides, and soil can accumulate As and subsequently manifest toxicity. Toxicity and phytoavailability vary greatly among plant species and soil conditions. Phytotoxicity followed the order green beans > lima beans = spinach > radish > tomato > cabbage (Woolson 1973), but this study covered only a limited number of species and soil conditions. Indeed, the experiment was performed under greenhouse conditions, which may be much different from field conditions. Woolson (1973) reported highest phytotoxicity for loamy sand, which can be reduced in field conditions by leaching As from the root zone.

The relationship between soil As and plant growth depends on its form and phytoavailability. Depression of rice growth seems to depend on the amount of As^{III} in soil, and arsines damage the roots of rice, resulting in inhibition of nutrient uptake (Takamatsu et al. 1983). Al–As and Ca–As fractions show significant adverse effects on rice growth and development (Li et al. 1985). Fruit trees grown on replanted orchard sites commonly exhibit retarded early growth, to which As toxicity may contribute (Davenport and Peryea 1991). Similarly, rice grown on former cotton-producing soils that had a history of repeated MSMA applications showed indications of susceptibility to straight head disease (abnormally developed or sterile flowers resulting in low grain yields) under flooded soils conditions (Wells and Gilmor 1977). Visual symptoms of phytotoxicity are only apparent at the highest application level of soluble As (60 mg/kg), when the youngest, most succulent tissue starts to wilt lightly and other tissues exhibit veinal necrosis (Tammes and De Lint 1969). Plants growing on soil with high As were stunted and displayed a red coloration (Farago et al. 2003). Greenhouse experiments showed that soils containing 72.1 mg/kg soluble As compounds were injurious to plant growth because higher-bioavailable As was used as liquid pesticide in orchard surface soil (Yan-Chu 1994).

Studies on As^V toxicity have shown that plant species that are not resistant to As suffer considerable stress on exposure, with symptoms ranging from inhibition of root growth to death (Macnair and Cumbes 1987; Meharg and Macnair 1991; Paliouris and Hutchinson 1991; Barrachina et al. 1995). As^V acts as a phosphate analogue and is transported across the plasma membrane via phosphate cotransport systems (Ullrich-Eberius et al. 1989). Once inside the cytoplasm, it competes with phosphate, for example, replacing phosphate in ATP to form unstable ADP-As, leading to the disruption of energy flow in the cell (Meharg 1994). As^{III} is also highly toxic to plants because it reacts with the −SH of enzymes and tissue proteins, leading to inhibition of cellular function and death (Ullrich-Eberius et al. 1989).

Organic As species are generally considered to be less toxic than inorganic species to a wide range of organisms, including aquatic plants, animals, and humans (Tamaki and Frankenberger 1992). It had been presumed that this was also true for terrestrial plants; however, research using a range of plants including *Spartina* sp., rice (*Oryza sativa*), and radish (*Rhapanus sativus* L.) has reported contradictory results. The phytoavailability of four As species in hydroponic systems to *Spartina patens* followed the trend DMA < MMA ~ As^V < As^{III}; however, the phytotoxicity was the reverse, with As^V ~ As^{III} < MMA < DMA. This result suggests that organic arsenicals were more toxic than inorganic species (Carbonell-Barrachina et al. 1998) and was confirmed in studies of radish where DMA was also identified as the most phytotoxic species (Carbonell-Barrachina et al. 1999; Tlustos et al. 1998). This higher phytotoxicity and phytoavailability of DMA and MMA was probably due to their greater upward translocation from root to shoot. It has been shown for rice that in a high-affinity uptake system (low substrate concentration), uptake of MMA and DMA is considerably less than that of As^V and As^{III} (Abedin et al. 2002a). Similarly, Tu and Ma (2002) reported that ladder brake (Chinese fern) was less effective in accumulating organic As forms than inorganic forms. These contradictory results suggest that the common hypothesis regarding inorganic species is more toxic than organic is true. However, organic species may be more toxic in a limited number of crops.

E. Exposure Pathway

Total exposure to As in human and animal will be the sum of exposures from the diet, drinking water, direct ingestion of soil and dust, inhalation, and percutaneous absorption. However, the pathway of exposure may vary from region to region depending on human activities, geological composition of the aquifer, and bioavailability of As (Grissom et al. 1999). Exposure by the general population occurs mainly through the ingestion of food and drinking water (NRC 1999; WHO 1981). Of these, food is generally the principal contributor to the daily intake of total As when water is As free. The predominant dietary source is seafood, followed by rice/rice cereals, mushrooms, and poultry (WHO 1998). However, the nature of As ingested in seafood is largely organic and therefore may not be toxic. In some areas, drinking water is a significant source of exposure to inorganic As. In these areas, drinking water often constitutes the principal contributor to daily intake. Contaminated soils such as mine tailings are also a potential source of exposure by dust inhalation and food (WHO 1998). People who produce or use As compounds occupationally, such as workers in nonferrous metal smelting, pesticide manufacturing or application, wood preservation, semiconductor manufacturing, or glass production, may be exposed to substantially higher As levels, mainly from dusts or aerosols in air. Human exposure pathways are shown in Fig. 7.

At least six groups of As compounds are present in the environment and can contribute to human exposure (WHO 1987):

1. Inorganic water-soluble compounds: oxides of As^{III} and As^V; soluble As^{III} and As^V salts
2. Inorganic compounds of low water solubility: some As^{III} and As^V salts, arsenides, arsenic selenide, and arsenic sulfide
3. Organic compounds occurring naturally or as pesticides, e.g., dimethylarsinic acid or cacodylic acid
4. Organic compounds occurring naturally in marine organisms, e.g., arsenobetaine, arsenocholine
5. Organic compounds used as feed additives, e.g., arsanilic acid
6. Gaseous inorganic and organic compounds, e.g., arsine

Terrestrial plants may accumulate As via root uptake from the soil or by foliar adsorption of airborne As deposited on the leaves. Levels are higher in biota collected near anthropogenic sources or in areas with geothermal activity. Some species accumulate substantial levels, with mean concentrations of up to 3000 mg/kg at arsenical mine sites (WHO 1998). Ma et al. (2001) discovered *Pteris vittate* (brake fern) as an As hyperaccumulating plant because of its high biomass production with high extraction from As-contaminated soils, but the mechanism involved for this hyperaccumulation is unclear. Tu and Ma (2002) suggested that plants take up As passively in conjunction with water flow at low soil concentration (<100 mg/kg). At higher soil concentrations, however, fern plants followed a different pathway. The behavior resembles immobile nutrient uptake, especially that of phosphorus; thus, they found higher As concentration in older fronds of ladder brake at moderate to high levels (up to 1100 mg/kg). A small number of tolerant plants has been recently discovered, of which water hyacinth (*Eichhornia crassipes*) is most promising for its As-absorbing capacity. It can remove up to 10 mg/L As from wastewater (Ingole and Bhole 2003) by means of absorption by its fibrous root system (Misbahuddin and Fariduddin 2002). The discovery of these As-accumulating plants has led to major research on the mechanism of uptake by plants with the sole aim to identify those plants that have the capacity to phytoremediate As-contaminated soils and water.

Arsenic in Food Chains Current literature suggests that As concentrations in the majority of crop plants are not a significant threat for human consumption

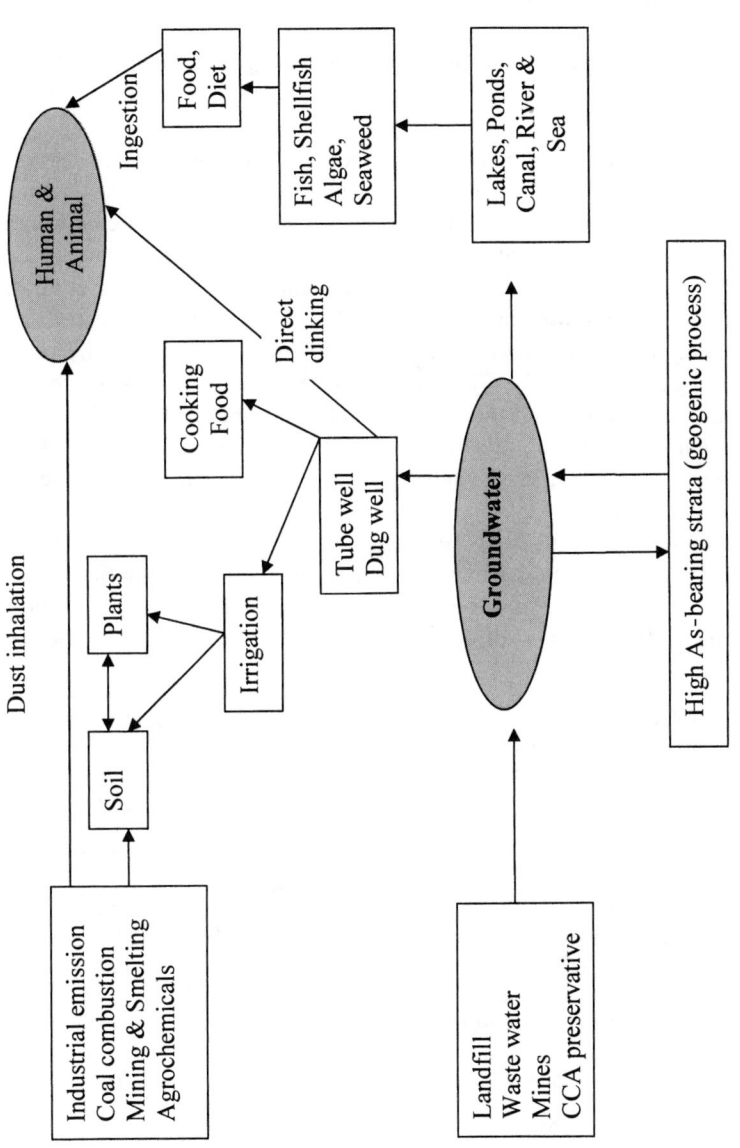

Fig. 7. Exposure pathways to arsenic.

(Lario et al. 2002; Alam et al. 2003; Lorenzini 2002; Munoz et al. 2002). However, vegetable and grain crops of Bangladesh and West Bengal are an exception, because a high As content is present in groundwater that has been used for agricultural irrigation for decades. Based on limited studies carried out in Bangladesh, high concentrations were found in vegetables and rice in affected areas, such as Chandipur village in Laximpur (Huq et al. 2001b). The highest concentration was recorded in arum root 20 mg/kg (dry weight). These data indicate that dietary habits may also contribute to the As problems in Bangladesh (Table 7).

The concentration of As in a range of food crops in control areas varied from <0.05 to >7 mg kg^{-1} dry weight (Huq et al. 2001a). These investigators also found the highest concentration in plant roots, indicating significant accumulation in the roots. Many leafy vegetables and herbs, in particular, have been reported to accumulate As. In West Bengal, India, Roychowdhury et al. (2002) investigated the total amount of As in food composites from affected areas of Murshidabad district and found the highest amount in arum leaf and rice, 0.33 and 0.23 µg/g, respectively. They also reported that As was absorbed by the skin of most vegetables and that fleshy vegetables contained significantly lower amounts. Chakraborti et al. (2001a) reported 95% and 5% inorganic and organic As species, respectively, in rice and 96% and 4% inorganic and organic respectively in vegetables.

Some controversy still exists regarding As concentration in Bangladesh rice grain. Some reports claimed that concentration in Bangladesh rice grains exceeded the safe limit, but most reports showed lower concentrations, which may be attributed to the widely different bioaccumulation potential of different rice

Table 7. Arsenic concentration (µg/g) in vegetables in an affected area of Bangladesh.[a]

Samples	Arsenic concentration (µg/g) d.w.		
	Min	Max	Average
Rice (*Oryza sativa*)	0.44	0.5	0.47
Arum, root (*Colocasia antiquorum*)	19.5	20.0	19.8
Arum, stem (*Colocasia antiquorum*)	0.74	0.92	0.83
Gourd, leaf (*Lagenoria siceraria*)	1.88	2.01	1.94
Brinjal (*Solanum melongena*)	0.18	0.20	0.19
Sweet potato, leaf (*Ipomea batatus*)	0.38	0.41	0.40
Pui shak (*Basilia alba*)	0.30	0.33	0.31
Kalmi shak (*Ipomea aquatica*)	0.29	0.35	0.32
Papaya (*Carica papaya*)	0.19	0.26	0.22
Red pumpkin (*Cucurbita maxima*)	0.20	0.27	0.24

[a]With kind permission from Mukherjee and Bhattacharya 2001, © 2001 NRC Canada.

cultivars. There is, however, a need for a detailed investigation of the effect of soil type and rice cultivars on bioaccumulation in rice grains. In some affected areas, rice grain contained as high as 1.7 mg/kg (Meharg and Rahman 2003), which suggests that a high As-containing paddy soil seasonally irrigated with contaminated water may produce rice grain that exceeds the food safety limit (1 mg/kg fresh weight). However, As concentration in rice straw, which is used as cattle feed in many countries including Bangladesh, was much higher, 91.8 mg/kg (Abedin et al. 2002b). High concentrations may have the potential for adverse health effects in cattle and could increase the risk of As exposure in humans via the plant–animal–human exposure pathway. However, exposure may vary depending on both the total and nature of As ingested in foods, with the effect being low from the ingestion of aquatic food (fish, prawns, etc.) that contain organic As compared to the more toxic inorganic forms. Given the limited information available on nature of As contained in food materials, total As derived from foods for a number of countries is given in Table 8. In Japan, higher values have been cited due to the presence of a higher percentage of organic arsenic in fish species.

V. Speciation of Arsenic
A. Definition of Speciation

There is no generally accepted definition of 'speciation,' and various workers have attributed various meanings to the term. Speciation may be defined as either (a) the process of identifying and quantifying the different, defined species, forms, or phases present in a material or (b) the description of the amounts

Table 8. Total arsenic (inorganic and organic) intake from foods in selected countries.

Country	As intake (µg As/d)	Year
Switzerland	30	1987
Holland	38	1989
Canada	49	1993
Finland	58	1980
Germany (West)	67	1983
England	89	1982
Denmark	118	1990
Japan	197	1990
PTWI[a]	130[b]	—

[a]PTWI, provisional tolerable weekly intake.
[b]Inorganic As intake per day is based on the weight of a person of 60 kg.
Source: With kind permission from Mukherjee and Bhattacharya 2001, © 2001 NRC Canada.

and kinds of these species, forms, or phases present (Ure et al. 1993a,b). Whichever definition is adopted, the species, forms, or phases are defined as (i) functionally, e.g., plant-available species, (ii) operationally, e.g., carbonate-bound species, or (iii) specific chemical compounds or oxidation states. Therefore, this definition focuses on the nature of interaction between chemicals and the substrate binding such chemicals. However, chemical speciation is concerned with studies of individual element species. Florence (1982) has defined speciation as "the determination of the concentration of the individual physico-chemical forms of a given element in a sample, which together constitute its total concentration." At IUPAC, Templeton et al. (2000) defined speciation analysis as the process leading to the identification and determination of the different chemical and physical forms of an element existing in a sample. Le (1999) broadly defined speciation as the identification and quantification of chemical species, their distribution and transformation in the environment, toxicity, and health effects. In this review, speciation is narrowly defined as separation, identification, and quantification of As^{III}, As^V, and organic As in a sample that are in available or soluble forms for most plants or animals. In this regard, our main focus is chemical speciation.

B. Need for Speciation

It is now well known that speciation is essential for understanding the distribution, mobility, toxicity, and bioavailability of chemical elements in natural systems. When evaluating interactions with the environment or to assess absorption, binding mechanisms, reactivity, and excretion of the element in humans, speciation can provide far more information than the analysis of elemental totals (Apostoli 1999). Kot and Namiesnik (2000) suggested that element speciation information is crucial for toxicity and biological activity of many elements, as these depend not only on their quantities but also on their oxidation states and/or chemical forms. For example, As is extremely toxic in its inorganic forms but relatively innocuous as arsenobetaine, a common form in fish.

Speciation of As is an important consideration as toxicity is mainly species dependent and is not well correlated with the total concentration. Therefore, determination of total As in a sample is of limited value because the result does not usually reflect the true level of hazard of that element. Hence, speciation is highly relevant in providing meaningful risk assessment data to assess the appropriate hazard level (Ng et al. 1998).

C. Speciation in Water, Soil, and Crops

As with most heavy metals and metalloids, there is a dearth of information on the nature of As species present in water, soil, and food crops. Speciation is critical given that human exposure pathways include each of these materials. Limited studies in Bangladesh show that As in drinking water is 100% bioavailable and may be present both in reduced (As^{III}) and oxidized from (As^V) (UNICEF 2001).

However, the ratio of As^{III} to As^V in tube-well water may vary depending on the sampling time. For instance, samples taken early in the morning show predominantly the reduced form while with progressing usage during the day there is significant transformation to the oxidized species (Naidu et al., unpublished data). However, an earlier study on speciation in groundwater reported As^V as the major species (Le 2002); this may be attributed to sample handling procedure. Speciation using disposable cartridges immediately after water sample collection addressed this limitation and demonstrated the presence of As^{III} in groundwater. These cartridges separate As^V from As^{III} by retaining As^V in the cartridge absorbent and allowed As^{III} to pass through to the filtrate. Total As can be determined from acidified water samples, and the As^V value is found from the difference between total As and that in the filtrate. Shraim et al. (2002) reported small amounts of DMA and MMA in tube-well water of West Bengal India. They also reported 60%–90% As^{III} and 20%–60% As^V in these samples. Their data were, however, based on only a limited number of samples. Similarly, Kim et al. (2002) studied groundwater samples from southeast Michigan and reported that most (53%–98%) of the detected species was arsenite As^{III}.

The nature of As species present in selected contaminated soils has been reported by Smith and Naidu (2004) for eight selected sites from South Australia, showing the presence of As^{III} up to 40% of the total in both surface and subsurface environment. However, these investigators did not provide reasons for the presence of As^{III} under alkaline and aerated surface soils.

As with soils and groundwater, there are few published studies on the nature of As in food crops. Heitkemper et al. (2001) investigated speciation in rice by ion chromatography and inductively coupled plasma mass spectrometry (IC-ICP-MS) and determined 11%–91% inorganic As of the total; DMA accounted for most of the remaining As in the samples. This result demonstrated that rice could be a threat for human consumption in contaminated areas as it contained high levels of inorganic As. Similarly, Schoof et al. (1999) reported 74%–110% inorganic As in raw rice in their market basket survey of inorganic As in food in the U.S. They also reported a small percentage of DMA and MMA in raw rice samples (Table 9). Abedin et al. (2002b) speciated rice straw instead of rice grain because of its high total As content and common use as cattle fodder. Trifluoroacetic acid (TFA) extracts of the sample indicated 72%–84% As^V, 15%–26% As^{III} and 1%–4% DMAA using anion-exchange HPLC–ICP-MS. These results suggest that high As^V content in rice straw should be avoided to minimize indirect exposure to humans. More speciation studies should be done on different food crops to obtain a clearer understanding of speciation, bioavailability, and toxicity relations to human, soil microbes, and crops. Most vegetables accumulate As in the peel; for example, garlic skin can accumulate nearly 100% inorganic As of the total (Munoz et al. 2002). Therefore, removal of vegetable peels before cooking can minimize As content in foodstuffs. Thus, it seems that without a clearer idea of metal species in soil, water, crops, and foodstuffs, it will be difficult to reliably assess risk.

Table 9. Inorganic arsenic species and their metabolites in environmental samples and food crops.

As species	Concentration range (% of total)	No. of samples analyzed	Matrix used	Reference
Arsenite	15–26	9	Rice straw	Abedin et al. 2002b
	32 (0.61 µg/g)	—	Whole carrot	Helgesen and Larsen 1998
	56	—	Radish root	Tlustos et al. 2002
	53–98	—	Groundwater	Kim et al. 2002
	60–90	9	Tube-well water	Shraim et al. 2002
Arsenate	>90 (<2 mg/L)	25	Soil solution	Naidu et al. 2000
	72–84	9	Rice straw	Abedin et al. 2002
	36 (0.67 µg/g)	—	Whole carrot	Helgesen and Larsen 1998
	40	—	Radish root	Tlustos et al. 2002
	20–60	9	Tube-well water	Shraim et al. 2002
Inorganic As	11–91	7	Polished rice	Heitkemper et al. 2001
	74–110	15	Raw rice	Schoof et al. 1999
	108	—	Garlic skin	Munoz et al. 2002
DMA	9–71	7	Polished rice	Heitkemper et al. 2001
	1–4	9	Rice straw	Abedin et al. 2002b
	13–91	15	Raw rice	Schoof et al. 1999
MMA	2–13	15	Raw rice	Schoof et al. 1999

D. Methods of Speciation

Speciation of As in real samples involves three major steps: extraction, derivitization and/or separation, and finally detection. Derivitization can be considered to be a combined separation and preconcentration method when the correct choice of derivitization conditions are used (Quevauviller et al. 1992). Hydride generation (HG) is the most common technique used for As derivatization (Quevauviller et al. 1992; Korte and Fernando 1991).

Although water samples may generally be examined directly, all solid samples must first be extracted before determination as the *in situ* determination of As compounds is currently impossible at environmental concentrations except with micro-XRD analysis. The extraction step should be mild to ensure that no transformation of species occurs during extraction and yet also highly efficient to ensure that all species present in a sample are extracted. Because no chemical extraction is currently ideal, a combination of various extractants is often required to reach this goal, and polar organic solvents or water are most commonly used (Goessler and Kuehnelt 2002).

Because of the different chemical properties of As compounds, a reliable separation within one single run is not possible. Therefore, a combination of various separation procedures must be employed. Hydride generation, liquid chromatography, gas chromatography, and capillary electrophoresis are commonly utilized.

Hydride Generation Technique (HG) HG was initially developed by Holak in 1969 and remains one of the most frequently used methods for arsenic determination at trace concentrations (Goessler and Kuehnelt 2001). It is based on the production of volatile arsines by either zinc/hydrochloric acid or sodium borohydride/acid mixtures. The volatile arsines are transported by an inert gas to the detection system.

Formation of arsine (AsH_3) from various As species usually involves the following steps (Shraim et al. 1999):

$$R_nAs(O)(OH)_{3-n} + H^+ + BH_4^- \rightarrow R_nAs(OH)_{3-n} + H_2O + BH_3 \quad (1)$$

$$R_nAs(OH)_{3-n} + (3-n)BH_4^- + (3-n)H^+ \rightarrow R_nAsH_{3-n} + (3-n)BH_3$$
$$+ (3-n)H_2O \quad (2)$$

$$BH_3 + 3H_2O \rightarrow H_3BO_3 + 3H_2 \quad (3)$$

The first reaction involves the reduction of As^V to As^{III}, and the arsines production occur in the second reaction.

The advantage of this method is that it can easily be connected to various detection systems *(AAS, ETAAS, ICP-AES, AFS, ICP-MS) and improves the detection limits as much as 100 fold over the commonly used liquid sample nebulization process (Goessler and Kuehnelt 2001). It can eliminate spectral and chemical interferences encountered in the detection system because only gaseous hydrides are introduced into the detector. However, some drawbacks have been reported (Howard 1997) for this method: (i) it is limited to the materials that form volatile arsines, (ii) the reaction conditions must be strictly controlled, and (iii) the presence of some interfering elements can reduce the efficiency of HG.

Liquid Chromatography (LC) Liquid chromatography is the most popular technique for As speciation in environmental samples. Most recent speciation studies in biological matrixes were performed by LC because of its both organic and inorganic species-determining capabilities. Among the various LC techniques, high performance liquid chromatography (HPLC), ion-exchange chromatography (IEC), and ion interaction chromatography (IIC) have frequently been used. In LC, a mobile phase is used to transport the sample and analyte into the column where individual species are selectively retained on the stationary phase and thus separated.

As^{III}, As^V, MMA, and DMA all form weak acids with significantly different dissociation constants (pKa). This difference in pKa values has been used to

---

*AAS = Atomic Absorption Spectrometry
ETAAS = Electrothermal Atomic Absorption Spectrometry
ICP-AES = Inductively coupled plasma—Atomic Emission
AFS = Atomic Floresence spectrometry
ICP-MS = Inductively coupled plasma—Mass Spectrometry

speciate acids by HPLC and IEC (Shraim 1999). The main advantage of this technique is that it removes the need for the complicated derivitization step and allows waters and extracts to be analyzed directly. LC possesses a wide range of applications, including complex organic matrices. Several types of chromatography can be used that make it easy to determine all As species, including inorganic and organic. However, coelution of species with similar physicochemical properties is a common problem in LC, and some organic solvents, that are used as the mobile phase have a limited UV transparency range, which limits their use with a UV detector. However, LC can be easily interfaced with many other detection systems such as ICP-MS, HG-AFS, and MS.

Gas Chromatography (GC) Gas chromatography is not commonly used for speciation because most As compounds are nonvolatile. Recently, determination of arsines, methylarsines, dimethylarsine, and trimethylarsine in air has been performed by GC-MS (Pantsar-Kallio and Korpela 2000). This method can be used to examine landfill gases for the determination of volatile arsenic compounds, but it is not applicable to biological samples.

Capillary Electrophoresis (CE) Capillary electrophoresis offers high separation efficiency and rapid analysis. It requires only nanoliters of samples, and the running costs are very low. During CE, As species are separated based on their charge-to-size ratio, which can be judiciously controlled by appropriate choice of buffer constituents and pH (Naidu et al. 2000). Currently, most CE separations are limited to pure standard solution or simple matrices. However, Naidu et al. (2000) speciated As in soil solution with capillary zone electrophoresis (CZE) and found As^V as the main species in aerated soil solution. However, their detection limit is 100 μg/L. This poor sensitivity is related to its short pathlength for UV detection, being limited by the size of the capillary used. Although some special capillaries are made with larger bubble pathlengths, connection to various detection systems is also difficult. Sample stacking techniques were also employed to improve CE sensitivity and were able to reduce detection limit to 25 μg/L in real water samples (Li and Li 1995).

The identification and quantification of As compounds is easily achieved with element selective detectors: ET-AAS, HG-FAAS*, AFS, ICP-AES, and ICP-MS. These detection systems have the advantage that only the As-containing compounds have to be separated. Of the various detection systems, AAS is the most widely used in arsenic speciation because of its sensitivity, simplicity, and precision at a low parts per billion (ppb) level (Korte and Fernando 1991). However, direct introduction of samples into the flame of the AAS was unsatisfactory because of matrix interference (Amran et al. 1995). ICP-MS is being increasingly used as a highly sensitive multielement analyzer for trace metal analysis of environmental and biological samples, although it is extremely ex-

---

*HG-FAAS = Hydride generation-flame atomic absorption spectrometry.

pensive and therefore inaccessible to many developing countries that have persistent As contamination issues.

Several combinations of derivitization, separation, and detection systems have been employed for arsenic speciation. The most common combinations are *CE-UV, HG-AAS, HPLC-ICP-MS, IC-ICP-MS, and HPLC-HG-AFS. Table 10 summarizes many of these hyphenated instrumental techniques.

E. Role of Speciation in Bioavailability

The role of chemical speciation is critical in mediating the bioavailability of most soil contaminants. Speciation studies give information on the distribution of chemical species of the element in a particular matrix by which one can easily understand the physicochemical properties of that species. For example, As^{III} is more mobile and soluble than As^V. Therefore, if it is possible to identify and quantify As^{III} and As^V in a sample, it will be clearer which species are more bioavailable. On the other hand, labile metal species are considered to be more biologically active than nonlabile fractions, and most of the bioavailability studies are generally based on the assumptions that greater solubility enhances bioavailability (Ng et al. 2003).

VI. Bioavailability

It is now well known that speciation of As plays an important role in determining both toxicity and bioavailability. Therefore, considerable controversy exists among toxicologists and regulators regarding the basis for existing guidelines for environmental and health risk assessments, because all current guidelines are based solely on total As concentrations. A metal can be toxic only if one or more of its toxic forms are bioavailable. Therefore, the maximum permissible limit should be based on the bioavailability of the specific toxic species; otherwise, the maximum permissible limit will be overestimated.

A. Definition of Bioavailability

The literature contains no precise definition of bioavailability. For a microbiologist, bioavailability is the concentration of a particular substance at a specific time that can cause an effect on the morphology or physiology of a specific organism (Naidu et al. 2001). For a plant scientist, bioavailability is the fraction that is available for the plant which can be termed as phytoavailability. In the context of toxicology, bioavailability refers to how much of a chemical is available to have an adverse effect on humans or other organisms (Kelly et al. 2002). In other words, bioavailability is the extent to which a chemical can be absorbed

---

*CE-UV = capillary electrophoresis-ultraviolet ray.
HPLC-ICP-MS = High performance liquid chromatography-inductively coupled plasma-mass spectrometry.

Table 10. Methods of arsenic speciation in environmental and biological samples.

Analytical techniques	Arsenic species and LOD (µg/L)		Application	Limitations	Reference
	As^{III}	As^{V}			
Hyphenated techniques of derivitization					
HPLC/UV-FI-HG-AAS	—	0.76	Urine (human)	Nonhydride-forming species require on-line digestion; As^{III}, As^{V}, MA coelute	Sur et al. 1999
HPLC-UV-HG-AFS	0.3	0.3	Fresh water, sewage waste, sediment Bivalves, bird egg	As^{III} and AsB coelute, nonhydride-forming species cannot be evaluated	Gomez-Ariza et al. 2000
LC-ICP-MS	0.15	0.11	Carrots (freeze-dried)	Peak broadening of As^{III} and As^{V} occurred	Vela et al. 2001
LC-UV-HG-AFS	1.2	1.0	Certified fresh water	Require both anion and cation exchange column to separate organic and inorganic As species	Vilano et al. 2000
LC-HG-ICP-MS	0.0005	0.001	Reference water sample	DMA cannot be determined expensive and complicated	Taniguch et al. 1999
HG-ICP-AES	1.0	1.0	Groundwater	Organic arsenic species can interfere	Muller 1999
HG-AAS	0.6	1.1	Aqueous solution and water	High concentration of acid interferes drastically for DMA determination, not suitable for complex matrices	Shraim et al. 1999
HG-ETAAS	0.075	—	Natural water (aqueous standard, hot spring water, seawater)	Reaction coil length must be optimized, hydride-forming medium should be optimum	Moreda-Pineiro et al. 2001

Table 10. (Continued).

Analytical techniques	Arsenic species and LOD (µg/L)		Application	Limitations	Reference
	As^{III}	As^V			
FI-HG-GFAAS	0.0015	0.0015	Seawater	Nonhydride-forming species need heat; difficult to obtain a similar response of all As species due to different analytical response	Carbon and Carbon 2000
FI-HG-AFS	0.11	0.07	Soil extract	Higher acid or alkali cause matrix effect, better sensitivity is pH dependent, As^V recovery is low	Shi et al. 2003
FI-EHG-AAS	0.4	0.6	Standard aqueous solution	Organic As species cannot be determined	Pyell et al. 1999
HG-ICP-MS	0.02	0.02	Drinking water	Nonhydride-forming As species cannot be evaluated	WCAS 2002
Hyphenated techniques of chromatography (HPLC)					
HPLC-ICP-MS	0.33	0.14	Human hair and nails	In fingernail, MMA^V, DMA^{III}, and iAs^{III} are poorly resolved	Mandal et al. 2003
HPLC-ICP-MS	0.07	0.09	Tube-well water	Very expensive, common laboratory cannot afford; >3000 ppm Cl^- interferes with As^V peak	Shraim et al. 2002
HPLC-ICP-MS	0.028	0.024	Lobster tissue extracts	As^V determination requires long time (e.g., 25 min); low recovery up to 80%, poor resolution between AsB and AsC	Brisbin et al. 2002
HPLC-ICP-MS	0.25	0.38	Natural water, soil extract, CRM	Expensive, difficult to prepare, handle and storage of sample hard due to redox processes	Guerin et al. 2000

Table 10. (Continued).

Analytical techniques	Arsenic species and LOD (µg/L)		Application	Limitations	Reference
	As^{III}	As^V			
HPLC-ICP-MS	0.03	0.04	Urine (normal, elevated)	Expensive, resolution between DMAA and As^{III} are poor, pH interference	Samanta et al. 2000
HPLC-ICP-MS	0.19	0.52	DORM-2 (dogfish muscle), mushroom (aq. extract)	Higher temperature causes peak splitting, matrix inference	Londesborough et al. 1999
HPLC-ICP-MS	0.03	0.05	Urine	$ArCl^+$ interferes with As^{III} peak, matrix interference such as 0.2% NaCl reduces As^{III} and DMA peak	Lintschinger et al. 1998
HPLC-ICP-MS	0.033	0.217	Contaminated soil, rainwater, soil pore water	Expensive, requires long time for As^V determination	Pongratz 1998
HPLC-ICP-MS	0.5	0.3	Water	—	Demesmay et al. 1994
HPLC-HG-AFS	0.05	0.06	Freshwater	Requires sample pretreatment, organic As species cannot be evaluated	Bohari et al. 2001
HPLC-HG-AFS	0.05	0.07	SRM (soil, sediment)	Nonhydride-forming species cannot be evaluated; limited use for biological samples	Gallardo et al. 2001
HPLC-HG-AFS	0.8	1.2	Urine	Requires optimization of buffer concentration and organic solvent	Ma and Lee 1998
HPLC-ICP-MS&ES-MS-MS	—	—	CRM (animal tissue); algae	Knowledge of the number of As species present in the sample is prerequisite	McSheehy et al. 2001
Hyphenated techniques of chromatography (IC)					
IC-ICP-MS	0.40	0.8	Water, soil	High cost	Vassileva et al. 2001
IC-ICP-MS	90	107	Acid mine draining water	pH interference	Gault et al. 2001

Table 10. (Continued).

Analytical techniques	Arsenic species and LOD (µg/L)		Application	Limitations	Reference
	As^III	As^V			
IC-ICP-MS	0.03	0.05	Fish, seafood, rice, edible mushrooms	Neutral cationic As species cannot resolve, RSD values are higher for lower concentration	Zbinden et al. 2000
IC-ICP-MS	0.64	2.19	Plant and soil extract	Organic As species are poorly resolved	Mattusch and Wennrich 2000
IC-ICP-MS	0.4	0.5	Water	Choice of the eluent for IC was found to be critical	Pantsar-Kallio and Manninen 1999
IC-ICP-MS	0.26	0.29	SRM, water	—	Mattusch and Wennrich 1998
IC-ICP-MS	0.10	0.30	Natural and waste water	High chlorine concentration caused column overloading and seriously affected the peak resolution	Terasahde et al. 1996
Hyphenated techniques of chromatography (CE)					
CE-ICP-MS	0.053	0.13	Aqueous solution	Controlling laminar flow, maintaining stable electrical connection to the interface	Day et al. 2000
CE-ICP-MS	15	15	Urine Sewage sludge (liquid)	High chlorine and high organic matter cause interferences	Mickalke and Schramel 1998
CE-ICP-MS	0.1	0.02	Water	The demonstrated direct insertion nebulizer (DIN) nebulization is poorer in detection limit than continuous nebulization; CE-DIN is commercially unavailable	Liu et al. 1995
CZE-ICP-MS	1.3	2.1	Drinking water, soil and humus extract	ArCl, laminar flow cause interference	Holderbeke et al. 1999

Table 10. (Continued).

Analytical techniques	Arsenic species and LOD (μg/L)		Application	Limitations	Reference
	As^{III}	As^{V}			
CZE-HG-ICP-MS	0.017	0.017 (Hyd)	Water	Peak broadening was observed; sample matrix effect	Magnuson et al. 1997
CE-MRBHG-ICP-MS	0.60	0.34	Standard solution	Complicated instrumental design	Tian et al. 1999
MRBHG-CZE-ICP-AES	320	320	Standard solution	Detection limit is poorer than other sample introduction mode	Tian et al. 1998
CZE-ESI-MS	500	1200	Liquid	Lack of sensitivity	Debusschere et al. 2000
LV-FASI-CZE	26	23	Pond water	Insensitive for complex matrix	Zhang et al. 2001
LV-FASI-CZE	800 12	3700 (Di) 25 (Stac.)	Natural (tap and spring water)	Relatively lower concentration sensitivity than HPLC or GC	Li and Li 1995
CZE-UV	90	60	Water, soil	—	Schlegel et al. 1996
	15	19	Standard solution	—	Amran et al. 1995
CE-InUV	5000	5000	Urine	Matrix effect; detection limit is very high	Wildman et al. 1991
CE-InUV	304 (Hyd) 124 (ELM)	89 (Hyd) 46 (ELM)	Water, soil extract	Lack of sensitivity and specificity; need sample stacking	Casiot et al. 1998
CE-InUV	500	100	Soil–water extract	Requires optimization of pH, capillary temperature, and run voltage	Naidu et al. 2000; Naidu 1996
CE-InUV	443	655	Coal fly ash extract	Requires optimizatin of pH, capillary temperature, and run voltage	Lin et al. 1995

LOD, limit of detection; EHG, electrochemical hydride generation; MRBHG, movable reduction bed hydride generation.

by a living organism and reach the systemic circulation (Kelley et al. 2002). Ng et al. (2003) defined bioavailability as the fraction of the element from an ingested matrix such as soil, water, or food that can be absorbed by an organism (e.g., humans). In general, bioavailability is the rate and extent of absorption and is a dynamic process.

Advances in evaluating the oral bioavailability of Pb and As have recently been critically reviewed by Ruby et al. (1999). They defined the term oral bioavailability as being that fraction of the administered dose that reached the blood from the gastrointestinal tract. The results from *in vitro* studies infer the oral bioaccessibility of a test material, bioaccessibility being defined as that fraction that is soluble in the gastrointestinal tract and is available for adsorption. The difference between bioavailability and bioaccessibility is therefore the difference between the amount actually absorbed into the bloodstream and that potentially available for absorption.

For human health risk assessment, absolute and relative bioavailability are the two important and separate measures. Absolute bioavailability is the fraction or percentage of a compound that is ingested, inhaled, or applied on the skin surface which is actually absorbed and reaches the systemic circulation (Hrudey et al. 1996). USEPA (1994) defined absolute bioavailability as the amount of a substance entering the blood via a particular route of exposure (e.g., gastrointestinal) divided by the total amount administered (e.g., soil lead ingested). Relative bioavailability is indexed by measuring the bioavailability of a particular substance relative to the bioavailability of a standardized reference material, such as soluble lead acetate (USEPA 1994) and is essential for environmental risk studies.

B. Factors Affecting Bioavailability

There are a number of soil, plant, and environmental factors that affect the bioavailability of As. Although soil chemical and environmental factors have been studied extensively (Yang et al. 2002; Turpeinen et al. 2003), limited information is available on how plant properties may influence As uptake and bioavailability.

Soil Factors Bioavailability of As in soil is strongly influenced by the nature and chemical and physical characteristics of soils including the nature of minerals and clay content, organic matter, texture, pH and Eh, cation-exchange capacity (CEC), and presence and concentration of oxides and hydroxides of Fe, Al, and Mn. The nature and proportion of soil constituents determine not only the extent but also the mechanism of adsorption and therefore the binding strength. Both As^{III} and As^{V} are strongly adsorbed to hydrous oxides of Fe, Mn, and Al in acid soils whereas calcium oxides in alkaline soils to a lesser extent adsorb anionic As species (Woolson et al. 1971). Anionic species are therefore in general more available to crops grown in alkaline soils (Frost and Griffin 1977). Generally, fine-grained soils limit the mobility of both As species more than coarse-grained soils as fine-grained soils contain large amounts of minerals and organic constituents. However, if the concentrations of iron and aluminium hydrous oxides in the soil are low, As tends to be mobile. Reduced bioavailability in soil is thought to be primarily a function of the presence of a less soluble mineral phase and ionic forms that are strongly adsorbed to soil particles or coprecipitated with other minerals. Factors that influence the adsorption capacity of soils influence the bioavailability and subsequent mobility of As in soils. More recent studies suggest that bioavailability in contaminated soils is also influenced by aging and sequestration (Turpeinen et al. 2003). Soil pH was the most important soil property affecting the decrease in bioaccessibility on aging (Yang et al. 2002).

Adsorption and Desorption Adsorption and desorption processes are the principal factors affecting the transport, degradation, and biological availability of compounds in soil (Smith et al. 1998). Hayes and Traina (1998) reported that As^V was strongly adsorbed on metal oxides and formed relatively insoluble precipitates with Fe and therefore became less bioavailable. They also reported that, at higher pH, As^{III} sorbed more weakly than As^V to metal oxides and therefore became bioavailable. In summary, soil conditions that promote precipitation or adsorption also tend to reduce the mobility and bioavailability of As. Numerous studies have demonstrated that the plant uptake of As is greatly influenced by its form in soil (Tu and Ma 2002; Burlo et al. 1999; Carbonell et al. 1998; Marin et al. 1992). The presence of other ions such as Fe and P also affects As availability and phytotoxicity (Fowler 1983).

Rhizosphere Interactions Rhizosphere interactions are deemed to play a key role in controlling bioavailability to crop plants (Fitz and Wenzel 2002). Plant roots modify the chemistry at the soil–root interface through the formation of organic substances and also through changes in hydrogen ion concentration (Naidu et al. 2003). Both changes can enhance the availability of nutrients and other ionic species. Root exudates may influence metal availability directly by acidification, chelation, precipitation, and oxidation and reduction reactions (Uren and Reisenauer 1988). However, virtually no information is available that directly addresses the fate of As in the soil–rhizosphere–plant system.

Plant Species Differences between plant species in accumulation of contaminants have been reported. Such variation may be attributed to the differences in tissue characteristics (e.g., leaf surfaces) and growth habits between species. A survey of 32 home gardens in an As mining area of Cornwall, UK, revealed the following differences between vegetable species in its accumulation in edible plant tissue (Thornton 1994):

Lettuce > onion, beetroot, carrot > peas, beans

Although Thornton (1994) reported greater accumulation of As in leafy vegetables compared to root vegetables, others (Merry et al. 1986) have reported greater accumulation in edible root tissue than edible leaf tissue. Such discrepancies may be due to the type of root vegetable considered and its preparation before analysis. Much of the As in root vegetables is contained in the outer skin (Woolson 1983a) and those that are consumed unpeeled may therefore be expected to exhibit higher tissue concentrations than those that are commonly peeled.

Plant species and relative abundance and availability of necessary elements also control metal uptake rates. Abundant bioavailable amounts of essential nutrients, such as P and Ca can decrease plant uptake of nonessential but chemically similar elements including As and Cd.

Microbial Processes Affecting Bioavailability Microorganisms influence the bioavailability of As directly by changing its chemical status and by modification of soil redox reactions. Various bacteria, fungi, and algae are able to oxidize and reduce As. The oxidation of reduced As species has been less widely studied than the processes of microbial reduction of As. The reduction of As^V to As^{III} is reportedly carried out by *Pseudomonas fluorescens* under anaerobic conditions, wine yeast, rumen bacteria, and cyanobacteria (Cullen and Reimer 1989). The reduction reaction is as follows:

$$As^{5+} + 2e^- = As^{3+}$$

Microbial methylation of As by common bacterial species in soil has been reported by a number of authors (Cullen and Reimer 1989; Bentley and Chasteen 2002). The bacterial methylation of inorganic As has been extensively studied in methanogenic bacteria. Methanogenic bacteria are a morphologically diverse group that produce methane as their primary metabolic end product under anaerobic conditions (Tamaraki and Frankenberger 1992). McBride and Wolfe (1971) reported that extracts of whole cells of the *Methanobacterium* strain MOH, growing anaerobically, reduced and methylated As^V to dimethylarsine. Dimethylarsine is only stable in the absence of O_2 and is rapidly oxidized under aerobic conditions (Cullen and Reimer 1989). In addition to bacteria, several fungi and algae are able to methylate As to the end product, trimethylarsine. Methylated forms of As mostly undergo volatilization loss from soils, and those remaining in the soil matrix are more mobile and known to be more toxic than nonmethylated forms (Bentley and Chasteen 2002).

C. Implications of Bioavailability to Toxicological Studies

From the standpoint of risk assessment at contaminated sites, proper assessment of bioavailability is critical to estimate potential exposure. As bioavailability may vary depending on the test species, the method used for assessment, and environmental conditions. This uncertainty in bioavailability can lead to erroneous estimates of the risk of exposure. Indeed, all the guidelines for human health risk assessment are based on only total metal concentrations. Therefore, considerable controversy exists between toxicologist and regulators. If maximum permissible concentrations (MPC) are based only on total metal concentration, MPC in soil or water sometimes violate the guidelines or exceed the limit. If permissible concentrations are based on bioavailability, then the MPC can be either equal to or less than the total concentration of metal. However, there is no certified method for measuring bioavailability, and thus unless better methods for bioavailability assessments are developed, total metal concentrations may be considered more conservative and perhaps a sound basis for risk assessment.

VII. Conclusions

Although recent published literature supports the hypothesis that toxicity and bioavailability vary with As species, few studies have considered the speciation of As and bioavailability at all. Thus, there is a great need to improve our knowledge of As speciation and bioavailability in a variety of environmental matrices.

Although hundreds of speciation methods have been developed throughout the world, cheap and matrix-sensitive techniques are still scarce. Moreover, a standard sample extraction technique is lacking, making intercomparison of As speciation studies from different groups difficult. Therefore, more effort needs to be focused toward developing and standardizing extraction methods, especially for green plant matrices.

Only limited information is available regarding the mechanism of As toxicity in plants, animals, and humans. The common As toxicity hypothesis is that As^{III} interferes with enzymes by bonding to $-SH$ and $-OH$ groups, and iAs^{V} as arsenate disrupts mitochondrial oxidative phosphorylation by competing with inorganic phosphate, but the mechanism and toxic effect of other inorganic and organic As species in humans and animals remain unclear. Therefore, there is an urgent need to direct studies toward the mechanism of As toxicity.

Contradictory information has been presented in the literature regarding the level of As in food crops. Some studies conducted on Bangladeshi crops suggest significant uptake of As depending on plant species and regional variation, whereas studies conducted elsewhere show that most plants do not accumulate As significantly in their edible parts. Moreover, species information for As in food crops and irrigation water is still lacking. Thus, more research should be carried out to ascertain As speciation variability with plant species, and also to assess its bioavailability from food crops, as well as toxicological information and guideline standards. Related to this, more effort needs to be directed toward the identification of As-tolerant crops that do not accumulate toxic As species. Scope also exists for studies on plant and microbial uptake of As and the possible use of As-hyperaccumulating plants for low-cost remediation of As-contaminated sites.

Summary

Although it is now commonly accepted that toxicity and bioavailability varies with As species, extensive research has been carried out on biological and environmental samples to assess toxicity and risk associated with As exposure based on total concentrations that may be in error. The health investigation guideline for the Australian environmental protection measure is 100 mg/kg (As_{tot}), which would cause potential risk to human health if all the As present in a sample were bioavailable (ANZECC 1992). Similarly, the MPC for As in food is 1 mg/kg (fresh weight), but this concentration may include contributions from As^{III}, As^V, and all organic species. Thus, a food substance, such as seafood, could have a high total concentration exceeding the guidelines, but most of the As would be in forms that are nontoxic to humans; i.e., the bioavailability is low, and the food would therefore be perfectly safe to eat. On the other hand, a food that has high bioavailability of As consequently is more toxic. Overall, it appears that contamination of water by As is probably more harmful to humans than As in food grains or vegetables, because As bioavailability in water is generally higher than its bioavailability in food. Nevertheless, As in food crops could make significant contribution toward total daily intake. Therefore, failure to consider the contribution of As species on their bioavailability could introduce a substantial bias into the estimation of risks associated with exposure as well as evaluation of As toxicity.

In conclusion, As must be regarded as an important environmental toxicant because of its acute and chronic toxic properties and extensive presence in the environment. Much remains to be learned about its toxicology and biochemistry for better understanding of this important contaminant.

Acknowledgments

Financial support from John Allwright Scholarship of the Australian Centre for International Agricultural Research (ACIAR) to K.F. Akter is acknowledged.

References

Abedin MJ, Feldmann J, Meharg AA (2002a) Uptake kinetics of arsenic species in rice plants. Plant Physiol 128:1120–1128.

Abedin MJ, Cresser MS, Meharg AA, Feldmann J, Howells JC (2002b) Arsenic accumulation and metabolism in rice (*Oryza sativa* L.). Environ Sci Technol 36(5):962–968.

Abernathy JR (1983) Role of arsenic chemicals in agriculture. In: Leaderer WH, Fentsterheim RJ (eds) Arsenic: Industrial, Biomedical, Environmental Perspectives. VNR, New York, pp 57–62.

Alam MGM, Snow ET, Tanaka A (2003) Arsenic and heavy metal contamination of vegetables grown in Samta village, Bangladesh. Sci Total Environ 308:83–96.

Amran B, Lagrade F, Leroy MJF, Lamotte A, Olle M, Albert M, Rauret G, L'opez-Sa'nchez JF (1995) Arsenic speciation in environmental matrices. Tech Instrum Anal Chem 17:185–304.

Andreae MO (1980) Arsenic in rain and the atmospheric mass balance of arsenic. J Geophys Res 85:4512—4518.

Andrew AS, Karagas MR, Hamilton JW (2003) Decreased DNA repair gene expression among individuals exposed to arsenic in United States drinking water. Int J Cancer 104:263–268.

ANZECC/NH&MRC (1992) Guidelines for the assessment and management of contaminated land, Canberra, Australia: Australia and New Zealand Environmental Council and National Health and Medical Research Council, Australia.

Apostoli P (1999) The role of element speciation in environmental and occupational medicine. Fresenius' J Anal Chem 363:499–504.

Armienta MA, Rodriguez R, Aguayo A, Ceniceros N, Villasenor G, Cruz O (1997) Arsenic contamination of groundwater at Zimapan, Mexico. Hydrogeol J 5:39–46.

Armienta MA, Brust H, Rodriguez R, Ongley LK, Ceneceros N, Aguayo A, Ceniceros N, Cruz O (2000) Clean water alternatives to arsenic naturally polluted groundwater in semiarid zone of Mexico. In: Bhattacharia P, Welch AH (eds) Arsenic in Groundwater of Sedimentary Aquifers. Precongress workshop abstract volume, 31st International Geological Congress, Rio de Janeiro, Brazil, pp 12–14.

Barrachina AC, Carbonell FB, Beneyto JM (1995) Arsenic uptake, distribution and accumulation in tomato plants—effect of arsenite on plant growth and yield. J Plant Nutr 18:1237–1250.

Basta NT, Rodriguez RR, Casteel SW (2001). Bioavailability and risk of arsenic exposure by the soil ingestion pathway. In: Frankenberger WT (ed) Environmental Chemistry of Arsenic. Dekker, New York, pp 117–138.

Belton JC, Benson NC, Hanna ML, Taylor RT (1985) J Environ Sci Health 20A:37.

Bentley R, Chasteen TG (2002) Microbial methylation of metalloids: arsenic, antimony, and bismuth. Microbiol Mol Biol Rev 66(2):250–271.

Berg M, Tran HC, Nguyen TC, Pham HV, Schertenleib R, Giger W (2001) Arsenic contamination of groundwater and drinking water in Vietnam: a human health threat. Environ Sci Technol 35:2621.

BGS/DPHE (2001) Arsenic contamination of groundwater in Bangladesh. In: Kinniburgh DG, Smedley PL (eds) British Geological Survey. Technical Report, WC/00/19, 4 vols. British Geological Survey, Keyworth.

BGS/MML (1999) Arsenic contamination of ground water in Bangladesh: a review. Prepared for the Department of Public Health Engineering, Government of Bangladesh, Dhaka, S5, p 54.

Bhattacharya P, Nordqvist S, Jacks G (1995) Status of As contamination in the soils around a former wood preservation facility at Konsterud, Kristinehamn Municipality, Varmland country, western Sweden. In: 5th Seminar on Hydrology and Environmental Geochemistry. Report 95,138. pp 70–72.

Bhattacharya P, Nordqvist S, Jacks G (1996) Heavy metals in soils: a case study for potential arsenic contamination in the environment around the site of former wood preservation facility in central Sweden. In: Aagaad P, Jahren J (eds) Report 69. The Rosenqvist Symposium on Clay Minerals in the Modern Society. Inst Geol. Univ. Oslo Norway, pp 9–10.

Bhattacharya P, Chatterjee D, Jacks G (1997) Occurrence of arsenic contaminated ground water in alluvial aquifers from delta plains, Eastern India: options for safe drinking water supply. Water Resour Dev 13:79–92.

Bhattacharya P, Jacks G, Frisbie SH, Smith E, Naidu R, Sarkar B (2002) Arsenic in the environment: a global perspective. In Sarkar B (ed) Handbook of Heavy Metals in the Environment. Dekker, New York, pp 147–215.

Bhumbla DK, Keefer RF (1994) Arsenic mobilization and bioavailability in soils. In: Nriagu JO (ed) Arsenic in the Environment, Part 1. Cycling and Characterization. Wiley, New York, pp 51–82.

Bissen M, Frimmel FH (2003) Arsenic: a review. Part I: Occurrence, toxicity, speciation, mobility. Acta Hydrochim Hydrobiol 31:9–18.

Bohari Y, Astruc A, Cloud J (2001) Improvements of Hg for the speciation of arsenic in natural fresh water samples by HPLC-HG-AFS. J Anal At Spectrom 16:774–778.

Borgono JM, Vicent P, Venturino H, Infante H (1977) Arsenic in the drinking water of the city of Antafagasta: epidemiological and clinical study before and after the installation of a treatment plant. Environ Health Perspect 19:103–105.

Brisbin JA, Bhymer C, Caruso JA (2002) A gradient anion exchange chromatographic method for the speciation of arsenic in lobster tissue extracts. Talanta 58:133–145.

Bundschuh J, Bonorino G, Viero AP, Albouy R, Fuertes A (2000) Arsenic and other trace elements in sedimentary aquifers in the Chaco-Pampean plain, Argentina: origin, distribution, speciation, social and economic consequences. In: Bhattacharia P, Welch AH (eds) Arsenic in Groundwater of Sedimentary Aquifers. Pre-Congress Workshop Abstract Volume. 31st International Geological Congress, Rio de Janeiro, Brazil, pp 27–32.

Burlo F, Guijarro J, Carbonell-Barrachina AA, Valero D, Martinez-Sanchez F (1999) Arsenic species: effects on and accumulation by tomato plants. J Agric Food Chem 47(3):1247–1253.

Carbon JY, Carbon N (2000) Determination of arsenic species in seawater by flow injection hydride generation in situ collection followed by graphite furnace atomic absorption spectrometry in stability of As (III). Anal Chim Acta 418:19–31.

Carbonell AA, Aarabi MA, Delaune RD, Gambrell RP, Patrick WH (1998) Arsenic in wet land vegetation: availability, phytotoxicity, uptake and effects on plant growth and nutrition. Sci Total Environ 217:189–199.

Carbonell-Barrachina AA, Aarabi MA, Laune De, Gambrell RD, Patrick WH (1998) The influence of arsenic chemical form and concentration on *Spartina patens* and *Spartina alterniflora* growth and tissue arsenic concentration. Plant Soil 198:33–43.

Carbonell-Barrachina AA, Burlo F, Lopez E, Martinez-Sanchez F (1999) Arsenic toxicity and accumulation in radish as affected by arsenic chemical speciation. J Environ Sci Health Part B Pesticides Food Contam Agric Wastes 34:661–679.

Casiot C, Alonso BCM, Boisson J, Olivier FXD, Potin-Gautier M (1998) Simultaneous speciation of arsenic, selenium, antimony and tellurium species in waters and soil extracts by capillary electrophoresis and UV detection. Analyst 123:2887–2893.

Cebrian ME, Albores A, Aguilar M, Blakely E (1983) Chronic arsenic poisoning in the north of Mexico. Hum Toxicol 2:121–133.

Chakraborti D, Rahman MM, Paul K, Saha KC, Quamruzzaman Q (2001a) Ground water arsenic contamination in South East Asia, with special reference to Bangladesh and West Bengal India. Presented at the International Workshop on Managing Arsenic in the South East Asia Region, 21–23 November, pp 1–4.

Chakraborti D, Basu GK, Biswas BK, Chowdhury UK, Rahman MM, Paul K, Chanda CR, Lodh D (2001b) Characterization of arsenic bearing sediments in Gangetic delta of West Bengal, India. In: Chappell WR, Abernathy CO, Calderon RL (eds) Arsenic Exposure and Health Effects. Elsevier, Amsterdam, pp 22–52.

Chappell J, Chriswell B, Olszowy H (1995) Speciation of As in a contaminated soil by solvent extraction. Talanta 42:323–329.

Chen CJ, Lin LJ (1994) Human carcinogenicity and atherogenicity induced by chronic exposure to inorganic arsenic. In: Nriagu JO (ed) Arsenic in the Environment. Part 11: Human Health and Ecosystem Effects. Wiley, New York, pp 109–132.

Chen CJ, Chiou HY, Huang WI, Chen SY, Hsueh YM, Tseng CH, Lin LJ, Lai MS

(1997) Systemic non-carcinogenic effects and developmental toxicity of inorganic arsenic. In: Chappell WR, Abernathy CO, Calderon RL (eds) Arsenic Exposure and Health Effects. Chapman & Hall, London, pp 124–134.

Chen SL, Dzeng SR, Yang MH, Chiu KH, Shieh GM, Wai CM (1994) Arsenic species in ground waters of the Blackfoot disease area, Taiwan. Environ Sci Technol 28: 877–881.

Chilvers DC, Peterson PJ (1987) Global cycling of As. In: Hutchinton TC, Meema KM (eds) Lead, Cadmium, Mercury and Arsenic in the Environment. Wiley, New York, pp 279–301.

Chiou HY, Hsueh YM, Liaw KF, Horng SF, Chiang MH, Pu YS, Lin JS, Huang CH, Chen CJ (1995) Incidence of internal cancers and ingested inorganic arsenic: a 7-year follow-up study in Taiwan. Cancer Res 55:1296–1300.

Choprapawon C, Rodcline A (1997) Chronic arsenic poisoning in Ronphibool Nakhon Sri Thammarat, the southern province of Thailand. In: Abernathy CO, Calderon RL, Chappell WR (eds) Arsenic Exposure and Health Effects. Chapman & Hall, London, pp 69–77.

Christie GL (1995) Speciation. In: Fifield FW, Haines PJ (eds) Environmental Analytical Chemistry. Chapman & Hall, London, p 279.

Clark PJ, Zingaro RA, Irgolac KJ, Mcginley AN (1980) Arsenic and selenium in Texas lignite. Int J Environ Anal Chem 7:295–314.

Clifford DA, Zhang Z (1993) Arsenic chemistry and speciation. In: Proceedings, Water Qual Technical Conference, Part II: Session 3D through ST6, pp 1955–1969.

Crecelius EA (1975) The geochemical cycle of arsenic in Lake Washington and its relation to other elements. Limnol Oceanogr 20(3):441–451.

Crecelius EA, Johnson CJ, Hoffer GC (1974) Contamination of soil near a copper smelter by arsenic, antimony and lead. Water, Air Soil Pollut 3:337–342.

Cullen WR, Reimer KJ (1989) Arsenic speciation in the environment. Chem Rev 98: 713–764.

Cuzick J, Sasieni P, Evans S (1992) Ingested arsenic, keratosis and bladder cancer. Am J Epidemiol 136:417–421.

Davenport JR, Peryea FJ (1991) Phosphate fertilizer influence leaching of lead and arsenic in soil contaminated with lead arsenate. Water Air Soil Pollut 57–58:101–110.

Day JA, Sutton KL, Soman SR, Caruso JA (2000) A comparison of capillary electrophoresis using indirect UV absorbance and ICP-MS detection with a self-aspirating nebulizer interface. Analyst 125:819–823.

Debusschere L, Demesmay C, Rocca JL (2000) Arsenic speciation by coupling capillary zone electrophoresis with mass spectrometry. Chromatographia 51:262–268.

Demesma C, Olle M, Porthault M (1994) Fresenius' J Anal Chem 348:205.

DG Environment (2000) Ambient air pollution by As, Cd, and Ni compounds. Position paper, working group on As, Cd and Ni compounds. DG Environment. European Commission, Belgium.

Dhar PK, Biswas BK, Samanta G, Mandal BK, Chakraborti D, Roy S, Jafar A, Islam A, Ara G, Kabir S, Khan AW, Ahmed SA, Hadi SA (1997) Groundwater arsenic calamity in Bangladesh. Curr Sci 73(1):48.

Dixon HBF (1997) The biochemical action of arsonic acids especially as phosphate analogues. Adv Inorg Chem 44:191–227.

Donohue JM, Abernathy CO (1999) Exposure to inorganic arsenic from fish and shellfish. In: Chappell WR, Abernathy CO, Calderon RL (eds) Arsenic Exposure and Health Effects. Chapman & Hall, London, pp 89–98.

DPHE/BGS/MML (1999) Groundwater studies for arsenic contamination in Bangladesh. Phase 1: Rapid investigation phase. BGS/MML Technical Report to Department for International Development, UK, vol 6.
Dudas MJ (1984) Enriched levels of arsenic in post-active acid sulfate soils in Alberta. Soil Sci Soc Am J 48:1451–1452.
Eisler R (1994) In: Nriagu JO (ed) Arsenic in the Environment, Part 11. Human Health and Ecosystem Effects. Wiley, New York, Chap. 11.
Ellenhorn MJ (1997) Ellenhorns' Medical Toxicology: Diagnosis and Treatment of Human Poisoning, 2nd ed. Williams & Wilkins, Baltimore, p 1540.
Farago ME, Kavanagh PJ, Leite M, Mossom J, Sawbridge J, Thornton G (2003) Uptake of arsenic by plants in southwest England. Biogeochemistry of environmentally important trace elements. Amer Chem Soc 835:115–127.
Ferguson JF, Gavis J (1972) A review of the arsenic cycle in natural waters. Water Res 6:1259–1274.
Fish RH, Brinkman FE, Jewett KL (1982) Fingerprinting inorganic arsenic and organoarsenic compounds in in situ oil shale retort and process water using a liquid chromatography coupled with an AAS as a detector. Environ Sci Technol 16:174–179.
Fitz WJ, Wenzel WW (2002) Arsenic transformations in the soil–rhizosphere-plant system: fundamentals and potential application to phytoremediation. J Biotechnol 99: 259–278.
Florence TM (1982) The speciation of trace elements in waters. Talanta 29:345–364.
Fowler B (1976) Environmental arsenic toxicity. In: Health Effect of Occupational Lead and Arsenic Exposure: A Symposium, Cincinnati, OH, pp 248–252.
Fowler BA (1977) Toxicology of environmental arsenic. In: Goyer RA, Mehlman MA (eds) Toxicology of Trace Elements. Wiley, New York, pp 79–122.
Fowler BA (1983) Biological and Environmental Effect of As. Elsevier, Amsterdam.
Frost RR, Griffin RA (1977) Effect of pH on adsorption of arsenic and selenium from land-fill leachate by clay minerals. Soil Sci Soc Am J 41:53–57.
Gallardo VM, Bohari Y, Astruc A, Potin-Gautier M, Astruc M (2001) Speciation analysis of arsenic in environmental solids. Reference materials by high-performance liquid chromatography—HG-AFS following orthophosphoric acid extraction. Anal Chim Acta 441:257–268.
Gault AG, Polya DA, Lythgoe PR (2001) Hyphenated IC-ICP-MS for the determination of arsenic speciation in acid mine drainage. Special Publication. Royal Society of Canada, Ottawa, pp 387–400.
Goessler W, Kuehnelt D (2001) Analytical methods for the determination of As and As compounds in the environment. In: Franjzenberzer WT (ed) Environmental Chemistry of Arsenic. Dekker, New York, pp 27–50.
Gomez-Ariza LJ, Sanchez-Rodas D, Giraldez I, Morales E (2000) A comparison between ICP-MS and AFS detection for arsenic speciation in environmental samples. Talanta 51:257–268.
Grantham DA, Jones FJ (1977) Arsenic contamination of water wells in Nova Scotia. J Am Water Works Assoc 69:653–657.
Grissom RE, Abernathy CO, Susten AS, Donohue JM (1999) Estimating total arsenic exposure in the United States. In: Chappell WR, Abernathy CD (eds) Arsenic Exposure and Health Effects. Elsevier, Amsterdam, pp 51–60.
Guerin T, Molenat N, Astruc A, Pinel R (2000) Arsenic speciation in some environmental samples: a comparative study of Hg-GC-QFAAS and HPLC-ICP-MS. Appl Organomet Chem 14:401–410.

Guha Mazumder DN, Gupta JD, Chakraborti AK, Chatterjee A, Das D, Chakraborti D (1992) Environmental pollution and chronic arsenicosis in south Calcutta. Bull WHO 70:481–485.

Gustafsson JP, Jacks G (1995) Arsenic geochemistry in forested soil profiles as revealed by solid phase studies. Appl Geochem 10:307–315.

Haswell SJ, O'Neill P, Bancroft CC (1985) Arsenic speciation in soil-pore waters from mineralized and unmineralized areas of south-west England. Talanta 32:69–72.

Hayes KF, Traina SJ (1998) Metal ion speciation and its significance in ecosystem health. Soil Chemistry and Ecosystem Health. Special publication no. 52. Soil Science Society of America, Madison, WI.

Heitkemper DT, Vela NP, Stewart KR, Westphal CS (2001) Determination of total and speciated arsenic in rice by ion chromatography and inductively coupled plasma mass spectrometry. J Anal At Spectrom 16:299–306.

Helgesen H, Larsen EH (1998) Bioavailability and speciation of arsenic in carrots grown in contaminated soil. Analyst 123:791–796.

Holak W (1969) Gas-sampling technique for arsenic determination by atomic absorption spectrometry. Anal Chem 41:1712–1713.

Holderbeke VM, Zhao Y, Vanhaecke F, Moens L, Dams R, Sandra P (1999) Speciation of six arsenic compounds using capillary electrophoresis-inductively coupled plasma mass spectrometry. J Anal At Spectrom 14:229–234.

Howard AG (1997) (Boro) Hydride techniques in trace element speciation. J Anal Atom Spectrom 12:267–272.

Hrudey SE, Chen W, Rousseaux CG (1996) Exposure routes and bioavailability factors for selecting contaminants. I. Arsenic and III. Chromium and chromium compounds. In: Hrudey SE (ed) Bioavailability in Environmental Risk Assessment. CRC Press, Boca Raton, FL.

Hung TC, Liao SM (1996) Arsenic species in the well water and sediments of the Blackfoot disease area in Taiwan. Toxicol Environ Chem 56:63–73.

Huq I, Smith E, Correll R, Ahmed A, Naidu R (2001a) Arsenic transfer in soil–water–crop environment in Bangladesh. 1. Assessing potential arsenic exposure pathways in Bangladesh. Presented at the International Workshop on Managing Arsenic in the South East Asia Region, 21–23 November, pp 50–51.

Huq SMI, Ara QAJ, Islam K, Zaher A, Naidu R (2001b) Possible contamination of As through food chain. In: Jacks G, Bhattacharya P, Khan AA (eds) Proceedings of the KTH-Dhaka University Seminar on Groundwater Arsenic Contamination in the Bengal Delta Plains of Bangladesh. KTH special publication. TRITA-AMI Report 3084, Stockholm, Sweden.

Ingole NW, Bhole AG (2003) Removal of heavy metals from aqueous solution by water hyacinth (*Eichhornia crassipes*). J Water Supply Res Technol-Aqua 52:119–128.

Jacks G, Bhattacharya P (1998) Arsenic contamination in the environment due to the use of CCA-wood preservatives. In: Arsenic in Wood Preservatives. Part 1. Kemi report 3/98, pp 7–75.

Jiang QQ, Singh BR (1994) Effect of different forms and sources of As on crop yield and arsenic concentration. Water Air Soil Pollut 74:321–343.

Johnstone RM (1963) Sulfhydryl agents: arsenicals. In: Hochster RM, Quastel JH (eds) Metabolic Inhibitors: A Comprehensive Treatise, vol 2. Academic Press, New York, pp 99–118.

Karim MM (2000) Arsenic in ground water and health problem in Bangladesh. Water Res 34(1):304–310.

Kelley ME, Brauning SE, Schoof RA, Ruby MV (2002) Assessing Oral Bioavailability of Metals in Soil. Battelle Press, Columbus, OH, pp 1–13.

Kim M, Nriagu J, Haack S (2002) Arsenic species and chemistry in groundwater of southeast Michigan. Environ Pollut 120(2):379–390.

Korte NE, Fernando Q (1991) A review of As(III) in groundwater. Crit Rev Environ Control 21:1–39.

Kot A, Namiesnik J (2000) The role of speciation in analytical chemistry. Trends Anal Chem 19(2-3):69–79.

Kumaresan M, Riyazuddin P (2001) Overview of speciation chemistry of arsenic. Curr Sci 80(7):837–846.

Lamb IA, Hughes MJ, Hughes CE (1996) Dispersion of As in soil and water of the Ballarat Goldfield, Victoria. In: First International Conference: Contaminants and the Soil Environment, Adelaide, Australia, pp 275–276.

Lario Y, Burlo F, Aracil P, Martinez-Romero D, Castillo S, Valero D, Carbonell-Barrachina AA (2002) Methylarsonic and dimethylarsinic acids toxicity and total arsenic accumulation in edible bush beans, *Phaseolus vulgaris*. Food Add Contam 19: 417–426.

Le CX (1999) Arsenic speciation in the environment. Can Chem News, pp 18–20.

Le CX (2002). Arsenic speciation in the environment and humans. In: Frankenberger WT (ed) Environmental chemistry of Arsenic. Dekker, New York, pp 95–116.

Leonard A (1991) Arsenic. In: Merian E (ed) Mineral and Their Compounds in the Environment: Occurrence, Analysis and Biological Relevance, 2nd ed. VCH, Weinheim, pp 751–773.

Li K, Li SFY (1995) Speciation of selenium and arsenic compounds in natural waters by capillary zone electrophoresis after on-column pre-concentration with field-amplified injection. Analyst 120:361–366.

Li SZ, Cui KZ, Wang LD (1985) Study of the fixed forms of arsenate and arsenite in soil polluted with arsenic in Dayu County, Jiangxi of China. Hebei Shifan Daxue Xuebao, Ziran Kexueban 1:61–70.

Lianfang W, Jianzhong H (1994) Chronic arsenism from drinking water in some areas of Xinjiang, China. In: Nriagu J (ed) Arsenic in the Environment, Part II: Human Health and Ecosystem Effects. Wiley, New York, pp 159–172.

Lin L, Wang J, Caruso J (1995) Arsenic speciation using capillary zone electrophoresis with indirect ultraviolet detection. J Chromatogr Sci 33:177–180.

Lin S, Cullen WR, Thomas DJ (1999) Methyarsenicals and arsinothiols are potent inhibitors of mouse liver thioredoxin reductase. Chem Res Toxicol 12:924–930.

Lintschinger J, Schramel P, Hatalak-Rauscher A, Wendler I, Michalke B (1998) A new method for the analysis of arsenic species in urine by using HPLC-ICP-MS. Fresenius' J Anal Chem 362:313–318.

Liu Y, Lopez-Avila U, Zhu JJ, Wiederin DR, Beckert WF (1995) Capillary electrophoresis coupled on-line with inductively coupled plasma mass spectrometry for element speciation. Anal Chem 67:2020–2025.

Liukkonen-Lilja H (1993) Arsenic in foods, Helsinki 1993. National Food Administration Research Notes 12, Helsinki, Finland, p. 16.

Loebenstein JR (1993) Arsenic. In: Mineral Yearbook 1990. U.S. Govt. Printing Office, Washington, DC, pp 167–170.

Londesborough S, Mattuch J, Wennrich R (1999) Separation of organic and inorganic arsenic species by HPLC-ICP-MS. Fresenius J Anal Chem 363:577–581.

Lorenzini G (2002) Trace elements in vegetables grown in an area exposed to the emissions of geothermal power plants. Fresenius' Environ Bull 11:137–142.

Luo ZD, Zhang YM, Ma L, Zhang GY, He X, Wilson R, Byrd DM, Griffiths JG, LaiLai S, He L, Grumski K, Lamm SH (1997) Chronic arsenicism and cancer in Inner Mongolia: consequences of well-water levels greater than 50 µg /L. In: Abernathy CO, Calderon RL, Chappell WR (eds) Arsenic Exposure and Health Effects. Chapman & Hall, London, pp 55–68.

Ma M, Le XC (1998) Effect of arsenosugar ingestion on urinary arsenic speciation. Clin Chem 44(3):539–550.

Ma LQ, Komar KM, Tu C, Zhang WH, Cai Y, Kennelley ED (2001) A fern that hyperaccumulates arsenic. Nature 409:579.

Macnair MR, Cumbes Q (1987) Evidence that arsenic tolerance in *Holcus lantus* L. is caused by an altered phosphate uptake system. New Phytol 107:387–394.

Maeda S (1994) Biotransformation of arsenic in the fresh water environment. In: Nriagu JO (ed) Arsenic in the Environment: Part 1, Cycling and Characterization. Wiley, New York, pp 155–187.

Magnuson ML, Creed JL, Brockhoff CAJ (1997) Speciation of arsenic compounds in drinking water by capillary electrophoresis with hydrodynamically modified electrophoretic flow detected through hydride generation inductively coupled plasma mass spectrometry with a membrane gas-liquid separator. J Anal At Spectrom 12:689–695.

Mandal BK, Suzuki KT (2002) Arsenic round the world: a review. Talanta 58:201–235.

Mandal BK, Roy Chowdhury T, Samanta G, Basu GK, Chowdhury PP, Chanda CR, Lodhi D, Karan NK, Dhar RK, Tamili DK, Das D, Saha KC, Chakraborti D (1996) Arsenic in ground water in seven districts of West Bengal, India: the biggest arsenic calamity in the world. Curr Sci 70(11):976–986.

Mandal BK, Ogra Y, Suzuki KT (2003) Speciation of arsenic in human nail and hair from arsenic-affected area by HPLC-inductively coupled argon plasma spectrometry. Toxicol Appl Pharmacol 189:73–83.

Marin AR, Masscheleyn PH, Patrick WH (1992) The influence of chemical form and concentration of arsenic on rice growth and tissue arsenic concentration. Plant Soil 139:175–183.

Marin AR, Masscheleyn PH, Patrick WH (1993) Soil redox-pH stability of arsenic species and its influence on arsenic uptake by rice. Plant Soil 52:245–253.

Masscheleyn PH, Delaune RD, Patrick WH (1991) Effect of redox potential and pH on arsenic speciation and solubility in a contaminated soil. Environ Sci Technol 25:1414–1418.

Mattusch J, Wennrich R (1998) Determination of anionic, neutral and cationic species of arsenic by ion chromatography with ICPMS detection in environmental samples. Anal Chem 70:3649–3655.

Mattusch J, Wennrich R (2000) Determination of arsenic species in water, soils and plants. Fresenius' J Anal Chem 366:200–203.

McBride BC, Wolfe RS (1971) Biosynthesis of dimethyl arsine by *Methanobacterium*. Biochemistry 10:4312–4317.

McDougall KW (1996) Arsenic and DDT residues at cattle Tick dip sites in NSW. In: First International Conference: Contaminants and the Soil Environment, Adelaide, Australia, pp 157–158.

McLaren RG, Naidu R, Tiller KG (1996) Fractionation of As in soils contaminated by cattle dip. In: First International Conference: Contaminants and the Soil Environment, Adelaide, Australia, pp 177–178.

McNeill LS, Chen HW, Edwards M (2001) Aspects of arsenic chemistry in relation to occurrence, health, and treatment. In: Franjzenberzer WT (ed) Environmental Chemistry of Arsenic. Dekker, New York, pp 141–153.

McSheehy S, Pohl P, Lobinski R, Szpunar J (2001) Complementary of multidimentional HPLC-ICP-MS and electrospray MS-MS for speciation analysis of arsenic in algae. Anal Chim Acta 440:3–16.

Meharg AA (1994) Integrated tolerance mechanisms: constitutive and adaptive plant responses to elevated metal concentrations in the environment. Plant Cell Environ 17: 989–993.

Mehrag AA, Macnair MR (1991) The mechanism of arsenate tolerance in *Deschampsia cespitosa* (L.), Beauv and *Agrostis* capillaries L. New Phytol 119:291–297.

Meharg AA, Rahman M (2003) Arsenic contamination of Bangladesh paddy field soils: implications for rice contribution to arsenic consumption. Environ Sci Technol 37(2): 229–234.

Merry RH, Tiller KG, Alston AM (1986) The effects of contamination of soil with copper, lead and arsenic on the growth and composition of plants. I. Effects of season, genotype, soil temperature and fertilizers. Plant Soil 91:115–128.

Michalke B, Schramel P (1998) Capillary electrophoresis interfaced to inductively coupled plasma mass spectrometry for element selective detection in arsenic speciation. Electrophoresis 19:2220–2225.

Misbahuddin M, Fariduddin A (2002) Water hyacinth removes arsenic from arsenic-contaminated drinking water. Arch Environ Health 57:516–518.

Moore JN, Ficklin WH, Johns C (1988) Partitioning of arsenic and metals in reducing sulfidic sediments. Environ Sci Technol 22:432–437.

Moreda-Pineiro J, Moscoso-Perez C, Lopez-Mahia P, Muniategui-Lorenzo S, Prada-Rodriguez D (2001) Talanta 53:871–883.

Mukherjee AB, Bhattacharya P (2001) Arsenic in ground water in the Bengal delta plain: slow poisoning in Bangladesh. Environ Rev 9:189–220.

Muller J (1999) Determination of inorganic arsenic (III) in ground water using hydride generation coupled to ICP-AES (HG-ICP-AES) under variable sodium boron hydride (NaBH$_4$) concentrations. Fresenius' J Anal Chem 363:572–576.

Munoz O, Diaz OP, Leyton I, Nunez N, Devesa V, Suner MA, Velez D, Montoro R (2002) Vegetables collected in the cultivated Andean area of northern Chile: total and inorganic arsenic contents in raw vegetables. J Agric Food Chem 50:642–647.

Murry RH, Tiller KG, Alston AM (1983) Accumulation of copper, lead and arsenic in some Australian orchard soils. Aust J Soil Res 21:549–561.

Naidu R (1996) Application of capillary electrophoretic analytical technique to anion speciation and analysis in natural systems. In: Abstract Book, ASSI and NZSSS National Soils Conference, pp 197–198.

Naidu R, Smith J, Mclaren RG, Stevens DP, Sumner ME, Jackson PE (2000) Application of capillary electrophoresis to anion speciation in soil water extracts: II. Arsenic. Soil Sci Soc Am J 64:122–128.

Naidu R, Megharaj M, Vig K, Kookana RS (2001) Bioavailability, definition and analytical techniques for assessment and remediation of contaminated (inorganic and organic) soils. Presented at the International Workshop on Chemical Bioavailability in the Terrestrial Environment, 18–20 November, Adelaide, South Australia, pp 4–6.

Naidu R, Gupta VVSR, Rogers S, Kookana RS, Bolan NS, Adriano DC (2003) Bioavailability, Toxicity and Risk Relationships in Ecosystems. Science Publishers, Enfield, NH, pp 41–57.

Nelson K (1983) Introduction: Industrial sources. In: Lederer WH, Fentsterheim RJ (eds) Arsenic: Industrial, Biomedical, and Environmental Perspectives. VNR, New York, p. 1.

NEPI (2000) Assessing the bioavailability of metals in soil for use in human health risk assessments. Bioavailability Policy Project Phase II: Metals Task Force Report. National Environmental Policy Institute, Washington, DC.

Ng CJ, Kratzmann SM, Qi L, Crawley H, Chiswell B, Moore MR (1998) Speciation and absolute bioavailability: risk assessment of arsenic-contaminated sites in a residential suburb in Canberra. Analyst 123:889–892.

Ng CJ, Barry N, Bruce S, Crawley H, Moore MR (2003) Bioavailability of metals and arsenic at contaminated sites from cattle dips, mined land and naturally occurring mineralization origins. In: Proceedings of the Fifth National Workshop on the Assessment of Site Contamination, EPHC, Adelaide, SA, pp 163–181.

Nickson RT, McArthur JM, Burgess WG, Ahmed KV, Ravencroft P, Rahman M (1998) Arsenic poisoning of Bangladesh groundwater. Nature (Lond) 395:338.

Nickson RT, McArthur JM, Ravenscroft P, Burgess WG, Ahmed KM (2000) Mechanism of arsenic release to groundwater, Bangladesh and West Bengal. Appl Geochem 15: 403–413.

NRC (National Research Council) (1977) Medical and Biologic Effects of Environmental Pollutants: Arsenic. National Academy of Sciences, Washington, DC.

NRC (1999) Arsenic in Drinking Water. National Academy Press, Washington, DC.

Nriagu JO, Pacyna JM (1988) Quantitative assessment of world wide contamination of air, water and soils by trace metals. Nature (Lond) 333:134–139.

Paliouris G, Hutchinson TC (1991) Arsenic, cobalt and nickel tolerances in two populations of *Silene vulgaris* (Moench) Garcke from Ontario, Canada. New Phytol 117: 449–459.

Pantsar-Kallio M, Korpela A (2000) Analysis of gaseous arsenic species and stability studies of arsine and trimethylarsine by gas chromatography-mass spectrometry. Anal Chim Acta 410:65–70.

Pantsar-Kallio M, Manninen PKG (1999) Optimizing ion chromatography: inductively coupled plasma mass spectrometry for speciation analysis of arsenic, chromium and bromine in water samples. In. J Environ Anal Chem 75(1-2):43–55.

Perker CL (1981) USEPA contract no 69-01-5965. The Mitre Corporation, XX, p 1.

Petrick JS, Jagadish B, Mash EA, Aposhian HV (2001) Monmethylarsonic acid and arsenite: LD_{50} in hamsters and in vitro inhibition of pyruvate dehydrogenase. Chem Res Toxicol 14:651–656.

Philips DJH (1994) The chemical forms of arsenic in aquatic organisms and their interrelationships. In: Nriagu JO (ed) Arsenic in the Environment, Part 1. Cycling and Characterization. Wiley, New York, Chap 11.

Pierce ML, Moore CM (1982) Adsorption of arsenite and arsenate on amorphous iron hydroxide. Water Resour 16:1247–1253.

Pongratz R (1998) Arsenic speciation in environmental samples of contaminated soil. Sci Total Environ 224:133–141.

Pyell U, Dworschak A, Nitschke F, Neidhart B (1999) Flow injection electrochemical hydride generation atomic absorption spectrometry (FI-EHG-AAS) as a simple device for the speciation of inorganic arsenic and selenium. Fresenius J Anal Chem 363: 495–498.

Quevauviller P, Donard OFX, Maier EA, Griepink B (1992) Improvement of speciation analyses in environmental matrices. Mikrochim Acta 109:169–190.

Ratnaike RN (2001) Arsenic in Health and Disease. Arsenic in the Asia—Pacific Region Workshop, Adelaide.
Reuther R (1992) Arsenic introduced into a littoral freshwater model ecosystem. Sci Total Environ 115:219–237.
Rimstidt JD, Chermak JA, Gagen PM (1994) Rates of reaction of galena, sphalerite, chalcopyrite and arsenopyrite with Fe (III) in acidic solutions. In: Alpers CN, Blowes DW (eds) Environmental Geochemistry of Sulphide Oxidation. American Chemical Society, Washington, DC, pp 2–13.
Ringwood K (1995) Arsenic in the gold and base-metal mining industry. Occasional paper no 4. Australian Minerals and Energy Environment Foundation (AMEEF), Melbourne, Australia, pp 1–33.
Rodriguez RR, Basta NT, Casteel SW, Pace LW (1999) An in vitro gastrointestinal method to estimate bioavailable arsenic in contaminated soils and solid media. Environ Sci Technol 33:642–649.
Rodriguez RR, Basta NT, Casteel SW, Armstrong FP, Ward DC (2003) Chemical extraction methods to assess bioavailable arsenic in soil and solid media. J Environ Qual 32:876–884.
Roy P, Saha A (2002) Metabolism and toxicity of arsenic: a human carcinogen. Curr Sci 82(1):38–45.
Roychowdhury T, Uchino T, Tokunaga H, Ando M (2002) Survey of arsenic in food composites from an arsenic-affected area of West Bengal, India. Food Chem Toxicol 40(11):1611–1621.
Russeva ED, Harvezo IP (1993) Speciation of As in natural waters: a review. Bulg Chem Commun 26:228–239.
Ruby MV, Schoof R, Brattin W, Goldade M, Post G, Harnois M, Mosby DE, Casteel SW, Berti W, Carpenter M, Edwards D, Cragin D, Chappell W (1999) Advances in evaluating the oral bioavailability of inorganics in soil for use in human health risk assessment. Environ Sci Technol 33(21):3697–3705.
Sadiq M, Alam I (1996) Arsenic chemistry in a groundwater aquifer from the eastern province of Saudi Arabia. Water Air Soil Pollut 89(1-2):67–76.
Samanta G, Chowdhury UK, Mandal BK, Chakraborti D, Sekaran NC, Tokunaga H, Ando M (2000) High performance liquid chromatography inductively coupled plasma mass spectrometry for speciation of arsenic compounds in urine. Microchem J 65: 113–127.
Schlegel V, Mattusch J, Wennrich V (1996) Speciation of arsenic and selenium compounds by capillary electrophoresis. Fresenius' J Anal Chem 354:535–539.
Schoof RA, Yost LJ, Eickhoff J, Crecelius EA, Cragin DW, Meacher DM, Menzel DB (1999) A market basket survey of inorganic arsenic in food. Food Chem Toxicol 37: 839–846.
Shi JB, Tang ZY, Jin ZX, Chi Q, He B (2003) Determination of As(III) and As(V) in soils using sequential extraction combined with flow injection hydride generation atomic fluorescence detection. Anal Chim Acta 477(1):139–147.
Shraim A (1999) Speciation of arsenic in environmental systems. PhD Thesis, University of Queensland, Australia, pp 27–28.
Shraim A, Chiswell B, Olszowy H (1999) Speciation of arsenic by hydride generation: atomic absorption spectrometry (HG-AAS) in hydrochlororic acid reaction medium. Talanta 50:1109–1127.
Shraim A, Sekaran NC, Anuradha CD, Hirano S (2002) Speciation of arsenic in tube-well water samples collected from West Bengal, India, by high-performance liquid

chromatography-inductively coupled plasma mass spectrometry. Appl Organomet Chem 16:202–209.
Smedley P (1996) Arsenic in rural groundwater in Ghana. J Afr Earth Sci 22(4):459–470.
Smedley PL, Kinniburgh DG (2002) A review of the source, behaviour and distribution of arsenic in natural waters. Appl Geochem 17(5):517–568.
Smedley P, Nocolli HB, Barros AJ, Tullio JO (1998) Origin and mobility of arsenic in groundwater from the Pampean plain, Argentina. In: Arehart GB, Hulston JR, Balkema AA (eds) Water–Rock Interaction. Rotterdam, Netherlands, pp 275–278.
Smedley PL, Edmunds WM, Pelig-Ba KB (1996) Environmental geochemistry and health. In: Appleton JD, Fuge R, McCall (eds), Geological Society Special Publication, vol 113. GJH, London, p 153.
Smith E, Naidu R (2004) Distribution and nature of arsenic along former railway corridors of South Australia. Science Total Environ (in press).
Smith E, Naidu R, Alston AM (1998) Arsenic in the soil environment: a review. Adv Agron 64:149–195.
Smith E, Naidu R, Alston AM (1999) Chemistry of arsenic in soils: I. Sorption of arsenate and arsenite by four Australian soils. J Environ Qual 28(6):1719–1726.
Smith E, Smith J, Smith L, Biswas T, Correll R, Naidu R (2003) Arsenic in Australian environment: an overview. J Environ Sci Health Part A Toxic/Hazard Subst Environ Eng 38:223–239.
Stater CS, Holmes HG, Byers HG (1937) U.S. Dept Agric Tech Bull 552:23.
Styblo M, Serves SV, Cullen WR, Thomas DJ (1997) Comparative inhibition of yeast glutathione reductase by arsenicals and arsenothiols. Chem Res Toxicol 10:27–33.
Sur RB, Begerow L, Dunemann, L (1999) Determination of arsenic species in human urine using HPLC with on-line phytooxidation or microwave-assisted oxidation combined with flow-injection HG-AAS. Fresenius' J Anal Chem 363:526–530.
Takamatsu T, Aoki H, Yoshida T (1983) Arsenic speciation in pot soil cropped with rice plant: fluctuations of arsenate, arsenite, momomethylarsonate and dimethylarsinate contents. Kokuritsu Kogai Kenkyusho Kenkyu Hokoku 47:153–163.
Tamaraki S, Frankenberger WT (1992) Environmental biochemistry of arsenic. In: Reviews of Environmental Contamination and Toxicology, vol 124. Springer-Verlag, New York, pp 79–110.
Tammes PML, De Lint MM (1969) Leaching of arsenic from soil. Neth J Agric Sci 17(2):128–182.
Taniguchi T, Tao H, Tominaga M, Miyazaki A (1999) Sentitive determination of three arsenic species in water by ion exclusion chromatography-hydride generation-inductively coupled plasma mass spectrometry. J Anal At Spectrom 14:651–655.
Templeton DM, Ariese F, Cornelis R, Danielsson L-G, Huntan H, Van Leeuwen HP (2000) Chemical speciation and fractionation of elements: definitions, structural aspects and methodological approach. Pure Appl Chem 72:1453–1470.
Terasahde P, Pantsar-Killio M, Manninen PKG (1996) Simultaneous determination of arsenic species by ion chromatography-inductively coupled plasma mass spectrometry. J Chromatogr A 750:83–88.
Thornton I (1994) Sources and pathways of arsenic in south-west England: health implications. In: Chappell W (ed) Arsenic Exposure and Health. Science and Technology Letters, Northwood, England, pp 61–70.
Thornton I, Farago M (1997) The geochemistry of arsenic. In: Abernathy CO, Calderon

RL, Chappell WR (eds) Arsenic Exposure and Health effects. Chapman & Hall, London, pp 1–16.

Tian X, Zhuang Z, Chen B, Wang X (1999) Determination of arsenic speciation by capillary electrophoresis and ICP-MS using a movable bed hydride generation systems. At Spectrum 20(4):127–133.

Tian XD, Zhuang ZX, Chen B, Wang XR (1998) Movable reduction bed hydride generation system as an interface for capillary zone electrophoresis and inductively coupled plasma atomic emission spectrometry for arsenic speciation analysis. Analyst 123: 899–903.

Tlustos P, Balik J, Szakova J, Pavlikova D (1998) The accumulation of arsenic in radish biomass when different forms of As were applied in the soil. Rostlinna Vyroba 44(1): 7–13.

Tlustos P, Goessler W, Szakova J, Balik J (2002) Arsenic compounds in leaves and roots of radish grown in soil treated by arsenite, arsenate and dimethylarsinic acid. Appl Organomet Chem 16:216–220.

Tseng WP (1977) Effects and dose-response relationships of skin cancer and blackfoot disease with arsenic. Environ Health Perspect 19:109–119.

Tu C, Ma LQ (2002) Effects of arsenic concentrations and forms on arsenic uptake by the hyperaccumulator ladder brake. J Environ Qual 31(2):641–647.

Turpeinen R, Virta M, Haggblom MM (2003) Analysis of arsenic bioavailability in contaminated soils. Environ Toxicol Chem 22(1):1–6.

Ullrich-Eberius CI, Sanz A, Novacky AJ (1989) Evaluation of arsenate and phosphate transport in *Lemna gibba* G1. J Exp Bot 40:119–128.

UN (United Nations) (2001) Synthesis report on arsenic in drinking water. In: Source and Behaviour of Arsenic in Natural Waters, pp 1–48. http://www.who.int/water_sanitaton_health/dwq/en/arsenicun1.pdf.

UNICEF (2001) Expert group meeting on arsenic, nutrition and food chain. UNICEF, New York, pp 1–53.

Ure A, Berrow M (1982) The elemental constituents of soils. In: Bowen HJM (ed) Environmental Chemistry. Royal Society of Chemistry, London, pp 94–203.

Ure AM, Quevauviller PH, Muntau H, Griepink B (1993a) Speciation of heavy metals in soils and sediments. An account of the environment and harmonization of extraction techniques under the auspices of the BCR of the Commission of the European Communities. Int J Environ Anal Chem 51:135–151.

Ure AM, Quevauviller PH, Muntau H, Griepink B (1993b) Improvements in the determination of extractable contents of trace metals in soil and sediment prior to certification. EUR Report 14763 EN. Commission of the European Communities, Luxembourg.

Uren NC, Reisenauer HM (1988) The role of root exudates in nutrient acquisition. Adv Plant Nutr 3:79–114.

USEPA (1994) Guidance manual for the integrated exposure, uptake biokinetic model for lead in children. EPA/540/R-93/081PB93-963510. U.S. Environmental Protection Agency, Washington, DC.

USEPA (2001). See website http://www.epa.gov/safewater/arsenic.html.

Varsanayi I, Zsofia F, Bartha A (1991) Arsenic in drinking water and mortality in southern great plain, Hungary. Environ Geochem Health 13:14–22.

Vassileva E, Becker A, Broekaert JAC (2001) Determination of arsenic and selenium species in groundwater and soil extracts by ion chromatography coupled to inductively coupled plasma mass spectrometry. Anal Chim Acta 441:135–146.

Vaughan GT (1993) The environmental chemistry and fate of arsenical pesticides in cattle dip sites and banana plantations. Investigation Report CET/LHIR 148. CSIRO, Divisions of Coal and Energy Technology, Centre for Advanced Analytical Chemistry, Sydney, Australia.

Vega L, Styblo M, Patterson R, Cullen W, Wang C, Germolec D (2001) Differential effects of trivalent and pentavalent arsenicals on cell proliferation and cytokine secretion in normal human epidermal keratinocytes. Toxicol Appl Pharm 172:225–232.

Vela NP, Heitkemper DT, Stewart KR (2001) Arsenic extraction and speciation in carrots using accelerated solvent extraction, liquid chromatography and plasma mass spectrometry. Analyst 126:1011–1017.

Vilano M, Padro A, Rubio R (2000) Coupled techniques based on liquid chromatography and atomic fluorescence detection for arsenic speciation. Anal Chim Acta 411:71–79.

Walsh LM, Sumner ME, Keeney R (1977) Occurrence and distribution of arsenic in soils and plants. Environ Health Perspect 19:67–71.

Webster JG (1999) Arsenic. In: Marshall CP, Fairbridge RW (eds) Encyclopaedia of Geochemistry. Chapman & Hall, London, pp 21–22.

Wei FS, Zheng CJ, Chen JS, Wu YY (1991) Huanjing Kexue 12:12.

Weiss CS, Parks EJ, Brinckman FE (1983) Speciation of As in fossil fuels and their conversion process fluids. In: Lederer WH, Fentsterheime RJ (eds) Arsenic: Industrial, Biomedical, Environmental Perspectives. VNR New York, pp 309–326.

Welch AH, Westjohn DB, Helsel DR, Wanty RB (2000) Arsenic in groundwater of the United States: occurrence and geochemistry. Ground Water 38(4):589–604.

Wells BR, Gilmor JT (1977) Sterility in rice cultivars as influenced by MSMA rate and water management. Agron J 69:451–454.

West Coast Analytical Service (WCAS), Canada (2002) Arsenic analysis and speciation by ICP-MS with hydride generation. http://www.wcas.com/TECH/ARSENIC.HTM.

Whetstone RR, Robinson WO, Byers HG (1942) US Dept Agric Tech Bull 797:32.

Whitecare RW, Pearse CS (1974) Arsenic and the environment. Miner Ind Bull 17:1–19.

WHO (1981) Arsenic. Environmental health criteria 18. WHO, Geneva.

WHO (1987) Arsenic. Air quality guidelines for Europe. WHO Regional Publications, European Series no 23. WHO Regional Office for Europe, Copenhagen.

WHO (1998) Arsenic Compounds. Environmental Health Criteria 224, 2nd ed. WHO, Geneva.

Wildman BJ, Jackson PE, William RJ, Peter GA (1991) Analysis of anion constituents of urine by inorganic capillary electrophoresis. J Chromatogr 546:459–466.

Woolson EA (1973) Arsenic phytotoxicity and uptake in six vegetable crops. Weed Sci 21:524–527.

Woolson EA (1983a) Emission, cycling and effect of As in soil ecosystems. In: Fowler BA (ed) Topics in Environmental Health: Biological and Environmental Effect of As. Elsevier, New York, pp 51–139.

Woolson EA (1983b) Industrial, Biomedical and Environmental Perspective. Van Nostrand Reinhold, New York.

Woolson EA, Axely JH, Kearney PC (1971) Correlation methods between available soil arsenic, estimated by six methods and response to corn (*Zea mays* L.). Soil Sci Am Proc 35:101–105.

Yan-Chu H (1994) Arsenic distribution in soils. In: Nriagu JO (ed) Arsenic in the Environment. Part 1: Cycling and Characterization. Wiley, New York, pp 17–49.

Yang J, Barnett MO, Jardine PM, Basta NT, Casteel SW (2002) Adsorption, sequestration and bioaccessibility of As(V) in soils. Environ Sci Technol 36(21):4562–4569.

Zbinden P, Andrey D, Blake C (2000) A routine ion chromatography ICP-MS method for the analysis of arsenic species applicable in the food industry. At Spectrom 21(6): 205–215.

Zhang P, Xu G, Xiong J, Zheng Y, Yang Q, Wei F (2001) Determination of arsenic species by capillary zone electrophoresis with large volume field amplified stacking injection. Electrophoresis 22:3567–3572.

Manuscript received March 30, accepted April 15, 2004.

Index

Acute toxicity, arsenic, 112
Adults blood lead levels, Brazil, 85, 87
Aldrin, marine mammals, 4
Analytical methods, arsenic, 122
Arsanilic acid, toxicity, 110
Arsenic acid (As^V), structure, 105
Arsenic acid, herbicide, 104
Arsenic, acute toxicity, 112
Arsenic, agricultural chemicals sources, 103
Arsenic, air content, 102
Arsenic, analytical methods food, 121
Arsenic, atmospheric levels, 102
Arsenic, ATP formation inhibition, 111
Arsenic, bioavailability, 120, 125
Arsenic, bioavailability, plant species, 132
Arsenic, bioavailability, rhizosphere interactions, 132
Arsenic, bioavailability, soil adsorption, 132
Arsenic, bioavailability, soil factors, 131
Arsenic, capillary electrophoresis speciation, 124
Arsenic chemistry, 104
Arsenic, chronic toxicity, 112
Arsenic, citric acid cycle effects, 110
Arsenic compounds, chemical structures, 105
Arsenic, derivitization methods (table), 126
Arsenic, drinking water guidelines, 114
Arsenic, energy metabolism inhibition, 110
Arsenic, environmental transfer pathways, 108, 109
Arsenic, enzyme systems inactivation, 110
Arsenic, EPA Maximum Permissible Conc drinking water, 101
Arsenic, food recommended guidelines, 114
Arsenic, gas chromatography speciation, 124
Arsenic, groundwater chemistry, 105
Arsenic, groundwater contamination, India, 97
Arsenic, groundwater levels, global, 102
Arsenic, human exposure pathway, 115, 117
Arsenic, human food intake (table), 119
Arsenic, hydride generation detection, 123
Arsenic, in food chains, 116
Arsenic, industrial sources, 103
Arsenic $LD_{50}s$, lab animals, 112
Arsenic, liquid chromatography speciation, 123
Arsenic, major compounds in human exposure, 116
Arsenic, mechanisms of toxicity, 110
Arsenic, modes of toxic action, 110, 113
Arsenic, oxidation states, 104
Arsenic oxyanions, 104
Arsenic, phytoxicity levels, 114
Arsenic, plant toxicity, 113
Arsenic, pyruvate oxidation inhibition, 110
Arsenic, Redox potential, groundwater (diagram), 106
Arsenic, removal from foods, methods, 121
Arsenic, seafood main dietary source, 115
Arsenic, seawater content, 101
Arsenic, smelting & mining sources, 103
Arsenic, soil chemistry, 107
Arsenic, soil content, global, 101
Arsenic, soil cycles, 108
Arsenic, soil speciation, 120
Arsenic, soil transformations (diagram), 108
Arsenic sources, anthropogenic, 103
Arsenic sources, environmental, 99
Arsenic sources, geogenic, 99
Arsenic sources, igneous rocks, 100
Arsenic sources, sedimentary rocks, 100
Arsenic sources, soils & sediments, 100
Arsenic speciation, defined, 119
Arsenic speciation in biological systems, 97 ff.

Arsenic, speciation methods, 122, 126
Arsenic speciation toxicity, 97 ff.
Arsenic, total intake from foods (table), 119
Arsenic, toxic effects humans & animals, 111
Arsenic, toxic effects plants, 113
Arsenic toxicity, 97 ff.
Arsenic toxicity, chemical forms, 109
Arsenic toxicity, species dependent, 120
Arsenic trioxide, major form industry-produced, 103
Arsenic, vegetable concentrations (table), 118
Arsenic, volatile forms, 108
Arsenic, water cycles, 108
Arsenic, water speciation, 120
Arsenic, WHO recommended value, drinking water, 101
Arsenic, wood preservative sources, 104
Arsenical herbicides, 104
Arsenical insecticides, 104
Arsenical pesticides, 104
Arsenicosis, pandemic Bangladesh, 97
Arsenious acid (As^{III}), structure, 105
Arsenobetaine (AsB), relative toxicity, 98, 105
Arsenocholine (AsC), relative toxicity, 98, 105
Arsenopyrite, arsenic-bearing mineral, 99
As^{-III}, arsine, 104
As^{III} (arsenious acid), relative toxicity, 98, 104, 107
AsB (arsenobetaine), relative toxicity, 98, 105
AsC (arsenocholine), relative toxicity, 98, 105
As^o, elemental arsenic, 104
As^V (arsenic acid), relative toxicity, 98, 104, 107
Atmospheric arsenic, 102
Atmospheric lead, Brazil, 69
Atmospheric lead, working environment limits, Brazil, 90
ATP formation inhibition, arsenic, 111

Baikal seals, epizootics, 2
Baleen whales, organohalogen contaminants, 2

Beluga whale blubber PHCs (table), 24
Beluga whales, 7
Bioaccumulation, lead Brazil, 61
Bioavailability, arsenic species, 125
Blood lead levels, adults, Brazil, 85, 87
Blood lead levels, chickens, Brazil, 79
Blood lead levels, children, Brazil, 83, 84
Blood lead levels, control populations, Brazil, 88
Blood lead levels vs. industrial sites, Brazil, 86
Blood lead levels vs. lead recycling plant, Brazil, 89
Blood lead levels vs. mining areas, Brazil, 87
Blood lead reference values, Brazil, 88
Body burdens, lead, Brazil, 81
Bottlenose dolphins, epizootics, 2
Brazil, air lead contamination regulations, 67
Brazil environmental lead contamination, 59 ff.
Brazil, lead ore reserve estimates, 60
Brazil lead problems, 59 ff.
Brazil, lead production, imports, exports, 60
Brazilian legislation, maximum lead permitted in foods, 77
By-catch delphinoid PHC studies, 35

Cacodylic acid, structure, 105
Calcium arsenate, insecticide, 104
California sea lions, reproductive impairment PHCs, 9
Capillary electrophoresis, arsenic speciation, 124
Caspian seals, epizootics, 2
Cetaceans, organohalogen contaminants, 1 ff.
Chemical structures, arsenic species, 105
Children's blood lead levels, Brazil, 83, 84
Chlordanes, marine mammals, 4
Chronic toxicity, arsenic, 112
Citric acid cycle effects, arsenic, 110
Cobaltite, arsenic-bearing mineral, 99
CYP isoenzymes, induction, marine mammals, 7

Cytochrome P450 induction, marine mammals, 7

DDT, marine mammals, 3, 4
Delphinapterus leucas, 7
Delphinoid blubber PBDEs (table), 28
Delphinoid blubber PCBs, hemispheres compared (fig.), 33
Delphinoid blubber PHCs, free-ranging (table), 37
Delphinoid blubber PHCs, *in vitro* assays, 39
Delphinoid liver, perfluorinated compounds (table), 26
Delphinoid populations, free-ranging, 35
Delphinoid studies, dead animal sampling, 35
Derivitization methods, arsenic (table), 126
Dieldrin, marine mammals, 4
Dimethylarsinic acid (DMA), relative toxicity, 98, 105, 107
Dioxins, marine mammals, 4
Disodium methanearsonate (DSMA), herbicide, 104
DMA (cacodylic acid), 105
DMA (dimethylarsinic acid), relative toxicity, 98, 107
Dolphin blubber PHCs (table), 12
Dolphin liver, perfluorinated compounds (table), 27
Dolphins, epizootics, 2
Dolphins, scientific names (table), 12
Drinking water, arsenic recommended guidelines, 114
Drinking water, maximum lead permitted, Brazil, 79, 80

Effluent water, maximum lead permitted, Brazil, 79
Emission regulation, air lead, Brazil, 67
Enargite, arsenic-bearing mineral, 99
Endocrine disruption, PHCs, marine mammals, 9
Endrin, marine mammals, 4
Environmental lead contamination, Brazil, 59 ff.

Ferric hydroxide, role in soil arsenic level, 107

Food, arsenic recommended guidelines, 114
Food, maximum permitted lead levels, Brazil, 77
Furans, marine mammals, 4

Gas chromatography, arsenic speciation, 124
Gasahol use, Brazil, 71
Gasoline ethanol content, Brazil, 71
Gasoline lead, air contamination, Brazil, 67
Gasoline lead content, Brazil, 71
Geogenic (natural) arsenic sources, 99
Global distillation, PHCs, 3
Global transport, PHCs, 3
Globicephala macrorhyncus, 7
Globicephala melas, 7
Grasshopper effect, PHCs, 3
Grey seals, epizootics, 2
Groundwater, arsenic content, global, 102

Halichoerus grypus, epizootics, 2
Halogenated dimethyl bipyrolles, global detection, 3
Harbour porpoises, 10
Harbour seals, epizootics, 2
Harp seals, 9
HCB (hexachlorobenzene), marine mammals, 3, 4
HCH (hexachlorocyclohexane), 4
HCH isomers, Arctic seawater concentrations, 4
Heptachlor, marine mammals, 4
Hexachlorocyclohexane (HCH), 4
Human exposure pathway, arsenic, 115, 117
Human lead exposure, Brazil, 59 ff.
Human lead exposure, Brazil, 77
Hunted delphinoid PHC studies, 35
Hydride generation, arsenic speciation, 123

Immunochemical assays, marine mammals, 7
In vitro assays, PHCs, delphinoids, 39
Industrial lead sites vs. blood levels, Brazil, 86
Isoenzymes (CYP), induction, marine mammals, 7

Killer whale blubber PHCs (table), 18
Killer whales, 11

LD_{50}s, arsenic lab animals, 112
Lead air contamination, Brazil, 66
Lead bioaccumulation, aquatic animals, Brazil, 62
Lead bioaccumulation, plants, Brazil, 62, 63
Lead, body burdens, Brazil, 81, 84
Lead, children's body burdens, Brazil, 82, 84
Lead contamination, aquatic organisms, Brazil, 65
Lead contamination, fish, Brazil, 64
Lead contamination, food containers, Brazil, 81
Lead contamination, maximum permitted in food, Brazil, 77
Lead contamination, school articles, Brazil, 81
Lead contamination, sediments, Brazil, 74, 76
Lead contamination, soil, Brazil, 70
Lead contamination sources, Brazil, 66
Lead contamination, surface waters, Brazil, 74
Lead contamination, underground water, Brazil, 70
Lead exports, Brazil, 62
Lead exposure, human, Brazil, 59, 77
Lead house dust contamination, Brazil, 66
Lead in urine, tetraethyl lead exposure indicator, 90
Lead, maximum permitted levels drinking water, Brazil, 79, 80
Lead, maximum permitted levels foods, Brazil, 77
Lead problems, Brazil, 59 ff.
Lead production, Brazil, 60
Lead, working environment limits, Brazil, 90
Leaded gasoline, air regulations, Brazil, 67
Liquid chromatography, arsenic speciation, 123

Magnesium arsenate, insecticide, 104
Marine mammal blubber PCBs, (figure), 29
Marine mammal tissue contaminants, 1 ff.
Marine mammals, halogenated contaminates, methods, 11
Marine mammals, pollutant accumulation pattern, 6
Marine mammals, pollutant biotransformation, 6
Marine mammals, reproductive impairment PHCs, 9
Maximum Permissible Concentration (EPA), arsenic drinking water, 101
Maximum permitted drinking water levels, Brazil, 79, 80
Maximum permitted food levels, lead, Brazil, 77
Mining lead sites vs. blood levels, Brazil, 87
Mirex, marine mammals, 4
MMA (monomethylarsonic acid), relative toxicity, 98, 107
MMAA (monomethylarsonic acid), soil mobility, 107
Mode of action, arsenic, 110
Monodontid blubber PHCs (table), 24
Monomethylarsine oxide, toxicity, 110
Monomethylarsonic acid (MMA), relative toxicity, 98, 107
Monosodium methanearsonate (MSMA), herbicide, 104
Mussel lead contamination, Brazil, 64
Mysticeti whales, 2

Narwhal blubber, PHCs (table), 25
Niccolite, arsenic-bearing mineral, 99

Odontoceti whales, 2
Organochlorine contaminants in whales, 1 ff.
Organohalogen contaminants in cetaceans, 1 ff.
Orpiment, arsenic-bearing mineral, 99
Oxyanions, arsenic, 104

Paris green, insecticide, 104
PBDEs, delphinoid blubber (table), 28
PBDEs, polybrominated diphenyl ethers, marine mammals, 3
PCBs, delphinoid blubber hemispheres compared (fig.), 33

Index

PCBs, dolphins, 2
PCBs marine mammal blubber (figure), 29
PCBs, marine mammals, 3
Perfluorinated compounds, delphinoid liver (table), 26
Persistent organohalogen contaminants (PHCs), physicochemical properties, 2
PFAs, perfluorinated acids, marine mammals, 3
PFOS, perfluorooctane sulfonates, marine mammals, 3
PHC studies, delphinoids, 11
PHCs (Persistent Organohalogen Contaminants), physicochemical properties, 2
PHCs, beluga whale blubber (table), 24
PHCs, delphinoid blubber, *in vitro* assays, 39
PHCs, dolphin blubber (table), 12
PHCs, free-ranging delphinoids blubber (table), 37
PHCs, global transport, 3
PHCs, "grasshopper" effect, 3
PHCs, killer whale blubber (table), 18
PHCs, monodontid blubber (table), 24
PHCs, narwhal blubber (table), 25
PHCs, phocoenid blubber (table), 20
PHCs, pilot whale blubber (table), 19
PHCs, porpoise blubber (table), 20
PHCs, sources & spatial distribution, 4
PHCs, spatial trends, marine mammals, 29, 36
PHCs, temporal trends, marine mammals, 34, 36
Phoca caspica, epizootics, 2
Phoca groenlandica, 9
Phoca hispida, 5
Phoca sibirica, epizootics, 2
Phoca vitulina, seal epizootics, 2
Phocoena phocoena, 10
Phocoenid blubber PHCs (table), 20
Physeter macrocephalus, 9
Pilot whale blubber PHCs (table), 19
Pilot whales, 7
Pinnipeds, organohalogen contaminants, 1ff.
Plasma retinol reduction, PHCs marine mammals, 10
Polychlorinated furans, marine mammals, 4
Porpoise blubber PHCs (table), 20
Porpoise liver, perfluorinated compounds (table), 26
Pyruvate oxidation inhibition, arsenic, 110

Redox potential, arsenic (diagram), 106
Reproductive impairment, PHCs, marine mammals, 9
Ringed seals, 5

Scientific names, dolphins (table), 12
Seafood, arsenic main dietary source, 115
Seals, harbour, epizootics, 2
Sediment contamination, lead, Brazil, 74, 76
Sediment lead criteria, Canada, 75
Soil contamination, lead, Brazil, 70, 72, 73
Soils, arsenic content, global, 101
Spatial trends, PHCs marine mammals, 29, 36
Speciation, defined, 120
Speciation methods, arsenic, 122, 126
Sperm whales, 9
Stenella coeruleoalba, epizootics, 2
Stockholm Convention on Persistent Organic Pollutants, 4
Stranded delphinoid PHC studies, 35
Striped dolphins, epizootics, 2
Study methods, marine mammals, contaminates, 11
Surface water contamination, lead, Brazil, 74
Surface water lead limits, Brazil, 75
Sururu mytella falcata, lead content, Brazil, 64

TCPM, tris(4-chlorophenyl)methane, 3
TCPMe, tris(4-chlorophenyl)methanol, 3
Temporal trends, PHCs marine mammals, 34, 36
Tetraethyl lead, air contamination, Brazil, 67
Tetraethyl lead exposure, lead in urine indicator, 90
Thyroid hormone reduction, PHCs marine mammals, 10

Tilapia rendalis, lead contamination, Brazil, 64
Toothed whales, organohalogen contaminants, 2
Toxaphene, marine mammals, 4
Tursiops truncatus, epizootics, 2

Underground water contamination, lead, Brazil, 70, 73
Urine, tetraethyl lead, human exposure indicator, 90

Water, arsenic content, global, 101
Whale halogenated contaminates, study methods, 11
Whales, organochlorine contaminants, 1 ff.
WHO recommended value, arsenic drinking water, 101

Zalophus californianus, 10